DATA ANALY EVERYONE

Turning Daily Life Into Data Insights

PUDHUVAI MURUGU

All inquiries should be addressed to
BitLit Publishing
10 Maybelle Court,
Mechanicsburg PA 17050, U.S.A
BitLit publishing is an imprint of TAMIL UNLIMITED LLC

For the latest on BitLit publishing please go to www.bitlitus.com.

For all enquiries contact@bitlitus.com Phone: 17177283999

Print ISBN: 979-8-9909905-7-9

Epub ISBN: 979-8-9909905-8-6

"To all who want to understand data

not just with tools, but with curiosity, courage, and care."

— *Pudhuvai Murugu*

Contents

Acknowledgements

Ever since I published my motivational book on Hesitation, I have been encouraged and inspired by friends, readers, and well-wishers who asked, "What's next?" Given my two decades of experience in the world of data, and with data becoming the foundation of modern innovation, from business intelligence to artificial intelligence, I received many requests to write a book that introduces **data from the very basics.**

This book is a response to that call. It is written for a **broad audience:**

Students who are curious about how data shapes our world

Working professionals from non-technical backgrounds who want to upgrade their skills and transition into data roles

Educators and trainers looking for simple ways to explain foundational data concepts

Parents who want to introduce their children to the data field

And even **curious learners** of any age who are exploring new career paths or knowledge areas.

I want to express my heartfelt gratitude to everyone who motivated me to write this book. Your encouragement gave me the strength to focus intensely and narrate each chapter in a clear, practical, and accessible way. I made every effort to avoid overwhelming readers with complex theories or technical jargon, choosing instead to build understanding step by step. If you're new to the data world, regardless of your background, I hope this book provides a confident and encouraging entry point. Whether you're a school student dreaming about the future, a professional exploring new horizons, or simply someone passionate about learning, this book is for you.

I welcome your questions, feedback, and suggestions. Please feel free to contact me via email. I look forward to hearing from you. I wish you all the very best as you begin your data journey.

Warm regards,

— *Pudhuvai Murugu*

saravanan.murugsv@gmail.com

Foreword

Sudarshan Meenakshi
Renal Engineering Technologist,
Toronto General Hospital
Faculty, Biomedical Engineering,
SETAS, Toronto
Part-Time Instructor,
Georgian College, Barrie,Toronto, Canada

meenakshi.sudarshan@uhn.ca

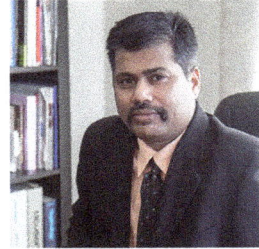

As someone who has spent years working in hospitals and biomedical education, I have consistently valued the importance of precision and evidence-based practices. Yet, until recently, I never consciously connected this work to "data analysis" in the formal sense. Reading Data Analysis for Everyone changed that for me.

What this book does so brilliantly is break down the idea that data analysis is only for IT professionals, statisticians, or coders. Through everyday examples from morning routines and grocery shopping to budgeting, fitness, and even emotional awareness, author Pudhuvai Murugu shows how we are all, in fact, data analysts in our own lives.

Each chapter in the book builds gently and thoughtfully, introducing practical ways to think analytically. In the early chapters, I was struck by how seamlessly the book connects daily decisions to concepts like pattern recognition, feedback loops, and productivity analysis. The later chapters delve into areas such as screen time, relationships, health metrics, and even civic data topics that felt surprisingly relevant to my work in healthcare and education.

As someone in the biomedical field, I found Chapter 7 on Body Data particularly compelling. The sections on health metrics, sleep tracking, and holistic well-being align closely with the principles we promote in clinical care. But what makes this book stand out is its accessibility; it doesn't rely on technical jargon or require any prior knowledge of data science. It's built on the idea that curiosity, observation, and reflection are the starting points.

Reading this book didn't just educate me, it motivated me. It motivated me to incorporate more data awareness into my teaching and patient care. I have started outlining a new book project that connects clinical technology with personal data habits, all inspired by the approachable framework Murugu has shared here.

Data Analysis for Everyone is more than a book. It's a mindset shift. Whether you're a student, parent, teacher, engineer, or healthcare professional, this book will help you see the patterns in your life more clearly and more confidently act on them.

Sudarshan Meenakshi

Introduction

In an era where data is widespread, understanding its role in our daily lives can transform the way we make decisions, solve problems, and interact with the world. "Data Analysis for Everyone" aims to demystify the concept of data analysis by illustrating how we interact with data in various aspects of everyday life, often without even realizing it. From choosing clothes based on the weather forecast to finding the quickest route to work using real-time traffic updates, our daily routines are increasingly driven by data.

For example, whenever I visit a grocery supercenter or a home repair store, instead of searching every aisle myself, I usually ask a floor representative where I can find the item I need. They often respond instantly, pointing me to the exact aisle and even the specific bin where the item is located. If it's not in their section, they quickly contact a specialist who knows precisely where it's stored.

This is very similar to how data queries work in the data world.

As a user, I make a "query" by asking for a specific item. The floor rep acts like a database index or query engine; they know the structure of the store (just like a database knows its schema) and can quickly fetch the location of what I need. If the request goes beyond their scope, they "join" with another data source (another rep) who specializes in that domain. In both cases, efficient retrieval depends on how well the system or store is organized and how quickly the query handler can access the correct location. Just as we retrieve the right data from the right table in seconds, I select the right product from the right shelf in moments.

If you can grasp the logic of this everyday scenario, you already hold a foundational understanding of data analysis. If not, this book will walk you through it, step by step. This book is written for those who assume that data analysis is the realm of IT professionals, data scientists, or math wizards. But the truth is, you don't need to be a programmer or statistician to make sense of data. This book demonstrates how everyday activities such as shopping, budgeting, or planning a trip are rooted in the same principles employed by data analysts worldwide.

By unpacking these familiar experiences, the book helps you learn how to think analytically, ask the right questions, and apply basic data concepts to enhance both personal and professional decisions. Through real-life analogies and clear explanations, each chapter aims to make data less intimidating and more empowering.

Each chapter explores various aspects of life, including health, finance, and relationships, illustrating how data analysis can improve understanding and efficiency. For instance, when managing personal finances, the book guides readers on how to track spending habits and make informed budgeting decisions, transforming financial anxiety into clarity and control. Similarly, in health and wellness, it examines how monitoring sleep patterns and exercise routines can lead to better lifestyle choices and improved overall well-being.

The book "Data Analysis for Everyone" is an invitation to view the world through a data-informed lens. It encourages curiosity and a proactive mindset, urging readers to ask the right questions, interpret patterns, and make decisions that are not only informed but also intentional. By the end of the book, readers will have a newfound appreciation for the data that surrounds them and the confidence to utilize it effectively in every aspect of their lives.

1.   You Already Analyze Data

You may not think of yourself as a data analyst, but you are one already. Every day, you gather information, recognize patterns, and make decisions, all without realizing it. Whether it's choosing what to wear based on the weather or adjusting your route due to traffic, your brain is doing real-time data analysis.

Understanding Everyday Decisions:

Data analysis is an ingrained part of our daily lives, often functioning beneath the surface of our awareness. It encompasses the collection of information, recognition of patterns, and execution of informed decisions based on this data. This process is not confined to the realm of professional analysts but is a fundamental aspect of everyday decision-making.

As consumers, we engage in a form of data analysis by comparing prices, checking expiration dates, estimating quantities based on past consumption, and evaluating discounts. Similarly, supermarkets engage in more complex data analysis. They track consumer behavior, including purchasing patterns and visit frequency, to optimize inventory, store layout, and promotional strategies. Both parties are involved in data analysis, albeit on different scales.

Grocery Shopping	Commute	Schedule	Cooking
Price Expiration	Traffic Time	Appointments	Preferences Time

Cooking dinner is another instance where analytical thinking is applied. A home cook assesses available ingredients, considers family preferences, and calculates the time available for meal preparation. Adjustments are made during cooking, such as altering the seasoning based on taste tests. These actions encompass resource evaluation, demand analysis, time management, and feedback application —core components of data analysis. A commuter might check traffic conditions via a navigation app, weighing route options based on real-time data and historical trends. In fashion,selecting an outfit involves a quick analysis of weather conditions, upcoming events, and previous attire choices

These scenarios demonstrate that data analysis is not limited to complex computations or specialized software. It is a natural, intuitive process that everyone engages in daily. The skills required—gathering information, recognizing patterns, estimating outcomes, and making decisions—are intrinsic to human behavior.

Recognizing our innate ability to analyze data empowers us to make more deliberate and informed choices. Whether it's optimizing a morning routine or making financial decisions, the principles of data analysis help us navigate life with greater clarity and control.

By acknowledging the role of data analysis in everyday life, individuals can develop a more analytical mindset. This involves being conscious of the data we encounter and actively applying it to decision-making. The goal is not to transform everyone into a data scientist, but to leverage the analytical skills we already possess to enhance our decision-making capabilities. This approach fosters a deeper understanding of how data shapes our lives, enabling us to harness its power more effectively.

Invisible Skills Made Visible

Analysis is an integral part of our everyday lives, often executed without conscious awareness. This invisible skill manifests itself in routine tasks, such as deciding what to cook, budgeting for groceries, or selecting the most efficient route to take for commuting.

Recognizing these skills is crucial to understanding how data analysis permeates our daily activities and decision-making processes.

Consider the act of cooking, where one evaluates available ingredients, the preferences of those being served, and the time constraints. This scenario involves inventory management and demand predictions, which are similar to business operations on a smaller scale. A cook intuitively analyzes data by recalling past meal outcomes, adjusting seasoning based on taste tests, and predicting future ingredient needs to minimize waste. Such decisions are driven by the same principles that guide formal data analysis: observation, pattern recognition, and optimization.

Similarly, commuting decisions are guided by data analysis. Checking traffic updates, assessing weather conditions, and evaluating historical travel times are all examples of using data to optimize routes and departure times.

This process involves real-time data assimilation and historical trend analysis, enabling individuals to make informed choices that reduce travel time and improve punctuality.

Provisioning is another domain where data analysis plays a pivotal role. Shoppers often compare prices, recall past consumption rates, and assess promotional offers to make informed, cost-effective purchasing decisions. This practice is comparable to a real-time analytics system, where historical data and current market conditions are synthesized to optimize shopping efficiency.

The act of dressing for the day also involves data analysis. Weather forecasts, the day's agenda, and recent clothing choices are taken into consideration to ensure comfort and appropriateness. This decision-making process highlights the use of forecasting, memory recall, and situational analysis to make data-driven wardrobe choices.

These examples demonstrate that data analysis is not confined to spreadsheets or algorithms; it is an inherent part of human cognition. By recognizing these everyday applications, individuals can enhance their analytical skills, becoming more intentional and effective in their actions.

The ability to consciously apply data analysis in daily life can lead to improved decision-making, greater efficiency, and a deeper understanding of how data shapes our experiences.

By making these invisible skills visible, individuals can harness the power of data analysis to transform their personal and professional lives. Whether it is through optimizing daily routines, making informed financial decisions, or boosting productivity, the principles of data analysis provide a framework for navigating the complexities of modern life with clarity and confidence. This chapter aims to empower readers to recognize and refine these skills, encouraging a more analytical perspective in everyday scenarios.

Instinctive Data Practices:

In our daily lives, we often make data-driven decisions without even realizing it. These actions are rooted in what can be termed instinctive data practices. These practices are informed by a mix of memory, pattern recognition, preferences, and predictions, forming the backbone of our everyday decision-making processes. This chapter explores how these instinctive practices manifest in various aspects of life and how they can be harnessed more consciously.

One of the most common areas where instinctive data practices are evident is in household management. Consider the simple act of meal planning. Without using any sophisticated tools, many people naturally conduct inventory analysis by checking the pantry, assessing the available ingredients, and recalling family preferences. This mental process involves forecasting and predicting meal sizes based on past consumption patterns, adjusting for expected guests, or considering upcoming events that might alter regular eating habits. Such instinctive practices ensure that meals are prepared efficiently and minimize waste.

Similarly, in personal grooming and fitness routines, instinctive data practices come to life. When brushing your teeth, you might notice your gums feel slightly tender. Instinctively, you might adjust your brushing technique, switch to a softer toothbrush, or try a different toothpaste. This is a form of anomaly detection where new data (gum sensitivity) prompts a behavior change based on past experiences and expected outcomes. In fitness, whether it's tracking steps, monitoring heart rates, or adjusting workout routines based on fatigue levels, individuals apply time-series data analysis to observe trends, identify gaps, and set future goals.

Transportation choices also illustrate instinctive data practices. When deciding whether to drive or take public transportation, individuals consider multiple factors, including the weather forecast, traffic patterns, personal schedules, and past experiences with delays or congestion. This multi-criteria decision-making process involves real-time data evaluation and historical trend analysis to optimize travel time and comfort.

Furthermore, instinctive data practices are crucial in financial management. Budgeting decisions often rely on recalling past expenses, anticipating future needs, and balancing priorities. Without realizing it, people engage in predictive analytics when they decide to save more during certain months or when they choose to cut back on discretionary spending after noticing a spike in expenses. These decisions are informed by a personal financial dashboard that tracks spending habits and income patterns over time.

In conclusion, instinctive data practices are an integral part of our daily lives, guiding decisions ranging from the mundane to the strategic. By becoming more aware of these practices, individuals can refine and enhance their decision-making capabilities. Developing a conscious understanding of how instinctive data practices work can lead to more intentional, efficient, and effective use of data in everyday life. This awareness transforms what might seem like routine actions into opportunities for improvement and optimization, demonstrating that everyone, in their way, is already a data analyst.

Forming a Data-Driven Mindset

Adopting a data-driven mindset is crucial for making informed and effective decisions. This process begins with an understanding that data is not merely a

FORMING A DATA-DRIVEN MINDSET

collection of numbers and statistics, but a tool that can significantly enhance our comprehension of the world around us. Developing this mindset involves shifting how we perceive data, not just as a component of technical fields but as a fundamental part of everyday decision-making.

To cultivate a data-driven mindset, one must first recognize the omnipresence of data in daily life. Whether it is through tracking personal expenses, analyzing time management, or observing dietary habits, data is continuously being generated and can be harnessed to improve individual and professional outcomes. The first step in this transformation involves becoming aware of the data we encounter daily and understanding its potential to inform decisions.

A crucial aspect of forming a data-driven mindset is learning to ask the right questions. This involves identifying what we want to know and why it matters. By framing clear, focused questions, we can gather relevant data and avoid the paralysis that comes with information overload. For instance, rather than vaguely wondering about productivity, one might ask, "What time of day am I most productive?" This specificity helps in collecting data that is actionable and useful.

Once the correct questions are framed, the next step is to collect and analyze data effectively. This doesn't always mean using sophisticated tools or software; sometimes, simple methods like maintaining a journal or using basic spreadsheet functions can provide valuable insights. The key is to start small and gradually build up to more complex analyses as skills and confidence grow.

Observation without judgment is another pillar of a data-driven mindset. It involves collecting data objectively and refraining from letting preconceived notions or biases skew interpretation. By observing patterns and outcomes without immediate evaluation, we allow the data to tell its story, guiding to more accurate insights and better decision-making.

Recognizing patterns is a natural progression in adopting this mindset. Once data is collected, identifying trends and correlations helps predict outcomes and inform decisions. This might involve recognizing that certain habits lead to specific results, such as noticing that a healthy breakfast correlates with increased morning productivity. Such insights can guide adjustments in behavior to enhance desired outcomes.

Experimentation and iteration are also integral to a data-driven approach. It involves testing hypotheses based on the data and making adjustments as necessary. This cycle of testing, observing, and refining is crucial for continuous improvement and adaptation. For instance, if data suggests that working out in the morning leads to better focus during the day, one might try adjusting their schedule to incorporate morning exercise and observe the results.

So, let me put it this way: forming a data-driven mindset is essentially about shifting how we perceive things in our daily lives. It's about asking the right questions, collecting and looking at data without bias, noticing patterns, and not being afraid to test things out. When we do that, we start making smarter decisions both at home and at work, everywhere. And the best part? It builds a habit of learning and improving all the time. That's the real power of thinking with data.

Try This – Reflect and Apply:

- Think of a decision you made today. What information did you use?
- Have you ever changed a routine (like what time you leave home) because of a pattern you noticed?
- Write down the steps you take when deciding what to cook. Which parts involve "data"?
- Over the next 3 days, observe how you make one repeated decision. What do you notice?
- Where in your daily life do you feel the most disorganized? Could tracking data help improve it?

2.   Thinking Like a Data Analyst

From Gut Feeling to Strategy:

In our daily lives, decisions often feel instinctual. We rely on gut feelings when choosing between options, whether it's what to eat for dinner or which route to take to work. Yet, beneath these seemingly spontaneous choices, there's a layer of subconscious data processing at play. This innate ability to analyze data is something everyone possesses, though we might not label it as such. Recognizing this skill is the first step towards harnessing it for strategic decision-making.

We make gut-based decisions all the time — choosing dinner, picking a route, managing our time. But beneath those instincts lies a hidden skill: **subconscious data processing.** We absorb patterns from experience, compare options, and respond. That's analysis, we don't call it that.

To move from instinct to insight, we need to:

Recognize the patterns in our daily decisions

Ask questions before assuming

Use past outcomes to refine future actions

For instance, don't just track your expenses. Categorize them. Spot your splurges. Adjust accordingly. That's how a data analyst thinks — and anyone can do it.

Core Functions of Analysis:

Let's look at what analysts do and how these habits apply to life:

- **Collect:** Notice what you track, money, time, energy, and mood.

- **Clean:** Filter out noise. Be honest. Remove excuses.

- **Explore:** Look for patterns. Are weekends more expensive? Mornings more productive?

- **Model:** Predict outcomes. If I skip breakfast, what happens?

- **Interpret:** Reflect. What does this say about your habits or goals?

- **Communicate:** Journal it. Talk about it. Share your insights.

These aren't technical steps, they're logical habits that can be practiced in small ways.

Data modeling is another pivotal function, where statistical and mathematical models are applied to the data to identify patterns and relationships. This step enables analysts to make predictions and draw conclusions about the dataset. Modeling can range from simple linear regressions to complex machine learning algorithms, depend ing on the complexity of the data and the questions posed.

Once models are developed, the following function is data interpretation. Analysts must translate the results into actionable insights. This involves understanding the implications of the data, assessing the reliability of the models, and considering the context in which the data exists. Practical interpretation requires not only technical skills but also domain knowledge to ensure that insights are relevant and applicable.

Ultimately, the communication of results is a vital component of data analysis, involving the clear and compelling presentation of findings through reports, dashboards, or presentations. The goal is to ensure that stakeholders can understand and act upon the insights provided. Effective communication requires the ability to simplify complex data and focus on the most pertinent information, tailoring the delivery to the audience's needs.

In summary, the core functions of data analysis, collection, cleaning, exploration, modeling, interpretation, and communication are integral to transforming data into valuable insights. Mastery of these functions empowers individuals and organizations to make informed decisions, optimize strategies, and finally achieve their objectives more effectively.

Frameworks and Tools:

In the realm of data analysis, frameworks and tools serve as the backbone for translating raw data into meaningful insights. These instruments are not mere accessories; they are essential components that streamline the data interpretation process, enabling analysts to make informed decisions efficiently and effectively.

Data frameworks provide structured approaches to data analysis, offering a blueprint for handling data systematically. These frameworks typically comprise a series of steps or phases that guide the analyst through the data collection, cleaning, analysis, and interpretation processes. A well-known framework in data analysis is the CRISP-DM (Cross-Industry Standard Process for Data Mining), which outlines a robust methodology involving business understanding, data understanding, data preparation, modeling, evaluation, and deployment. Such frameworks help ensure that the analysis is thorough and that all necessary aspects of the data are considered.

When it comes to tools, the selection is vast and varied, each offering unique functionalities that cater to different aspects of data analysis. Spreadsheets, such as Microsoft Excel and Google Sheets, remain popular due to their accessibility and versatility. They allow users to perform basic data manipulations, create pivot tables, and visualize data through charts and graphs. These tools are handy for those new to data analysis or for handling smaller datasets.

For more sophisticated data manipulation and analysis, programming languages like Python and R are indispensable. Python, with libraries such as Pandas, NumPy, and Matplotlib, offers powerful capabilities for data manipulation, statistical analysis, and visualization. R is particularly favored in statistical circles for its extensive range of packages tailored for various types of data analysis, from linear modeling to time series analysis.

Database management systems, such as SQL (Structured Query Language), are crucial for handling large datasets. SQL allows analysts to query databases efficiently, extracting and manipulating data stored in relational database systems. It is a fundamental tool for anyone working with large volumes of data, as it enables the retrieval of specific data subsets necessary for analysis and interpretation.

Visualization tools, such as Tableau, Power BI, Looker, Qlik Sense, and SAS Analytics, transform raw data into interactive and intuitive visual formats. These tools are essential for communicating insights clearly and effectively, allowing analysts to create dashboards that provide stakeholders with a visual representation of data trends and patterns. The ability to drill down into data through interactive visuals can reveal insights that are not immediately apparent through raw data alone.

In addition to these, machine learning tools and platforms have become integral to modern data analysis. Tools such as TensorFlow and Scikit-learn enable analysts to build predictive models that can identify trends and forecast future outcomes based on historical data. These tools harness the power of algorithms to uncover complex patterns within datasets, providing a deeper level of insight.

The integration of frameworks and tools in data analysis is not just about using the latest technology but about selecting the right combination that aligns with the specific needs of a project. The choice of framework and tools can significantly impact the efficiency and accuracy of the analysis, making it crucial for analysts to be knowledgeable about the options available and their respective strengths and limitations. Through a strategic selection of frameworks and tools, analysts can ensure that they are leveraging data to its fullest potential, driving informed decision-making, and delivering valuable insights. The more we understand our data, the easier it becomes to explore different tools. We don't have to learn everything at once; we need to pick up what we need when we need it. Over time, it all adds up, helping us work smarter.

Mindset Shifts:

The transformation from relying on assumptions to focusing on inquiry is a pivotal shift in data analysis. This change involves moving away from vague feelings of busyness or lack of motivation and beginning to ask precise questions about one's

daily activities, energy levels, and expenditure. For instance, rather than lamenting about feeling busy without accomplishment, the new approach encourages identifying the top time-consuming activities each day. Similarly, questioning one's energy peaks during the day can replace a general sense of demotivation.

This kind of questioning lays the foundation for more structured data gathering and analysis.

To facilitate this shift, start by choosing an area of life you wish to enhance, such as time management or financial planning. Develop a central question, identify data sources, and propose a small, actionable step. For example, if time management is the focus, one might ask, "Where do I lose time?" and use a daily calendar or app usage data to log tasks over a few days. The emphasis is not on perfect data collection but on beginning the process of observation and inquiry.

Transitioning from a thought-based to an action-based mindset is crucial in data analysis. This involves breaking down problems and testing ideas quickly rather than overthinking them. The process consists of cultivating curiosity, posing more effective questions, observing the environment, tracking feedback, and adjusting strategies based on the results. This iterative approach is mirrored in everyday activities, such as cooking, where meal planning involves checking ingredients, considering family preferences, and adjusting recipes based on feedback. Such practices illustrate how data-driven decisions are often made instinctively, without the use of formal tools.

In essence, the key is to recognize that data is prevalent in everyday life, from inventory checks in the kitchen to planning meal quantities based on family activities and schedules. These activities are a form of data analysis, involving memory, pattern recognition, and prediction. For instance, noticing that certain groceries spoil before use can lead to purchasing smaller quantities or adjusting meal plans to incorporate ingredients sooner. These minor adaptations are examples of feedback-driven improvements.

Essentially, the mindset shift in data analysis involves adopting a proactive and inquisitive approach to life's challenges. By framing everyday decisions as data-driven processes, one becomes more adept at identifying patterns, making informed adjustments, and achieving better outcomes. The focus is on continuous learning and improvement, utilizing available data to make informed choices. This mindset not only enhances personal efficiency but also prepares individuals for broader applications of data analysis in professional and community contexts.

3. Routines as Datasets

You may have heard the word "dataset" and thought it belonged only in the world of computers or science. But in reality, you create and use datasets every single day, without even realizing it. A dataset is simply a collection of information. It could be a list of items, times, choices, feelings, or actions. When these are collected over time and organized, they become a powerful tool for identifying patterns and making informed decisions.

Mornings and Movement:

Your day begins with a set of routines that may seem normal but reveal a great deal. These morning habits aren't just chores; they're patterns filled with valuable data. From the moment you wake up, your actions follow habits you've built over time. Noticing these patterns can help you improve your routine, make mornings smoother, and start your day with less stress. The initial hour of your day can be viewed as a micro-workflow, consisting of several repeatable actions, such as waking up, personal hygiene, dressing, eating breakfast, and transitioning to work. Each of these actions is a data point that can be observed and adjusted to improve your daily flow. If you find yourself consistently rushing or feeling stressed, there is likely a breakdown within this sequence that can be identified and addressed.

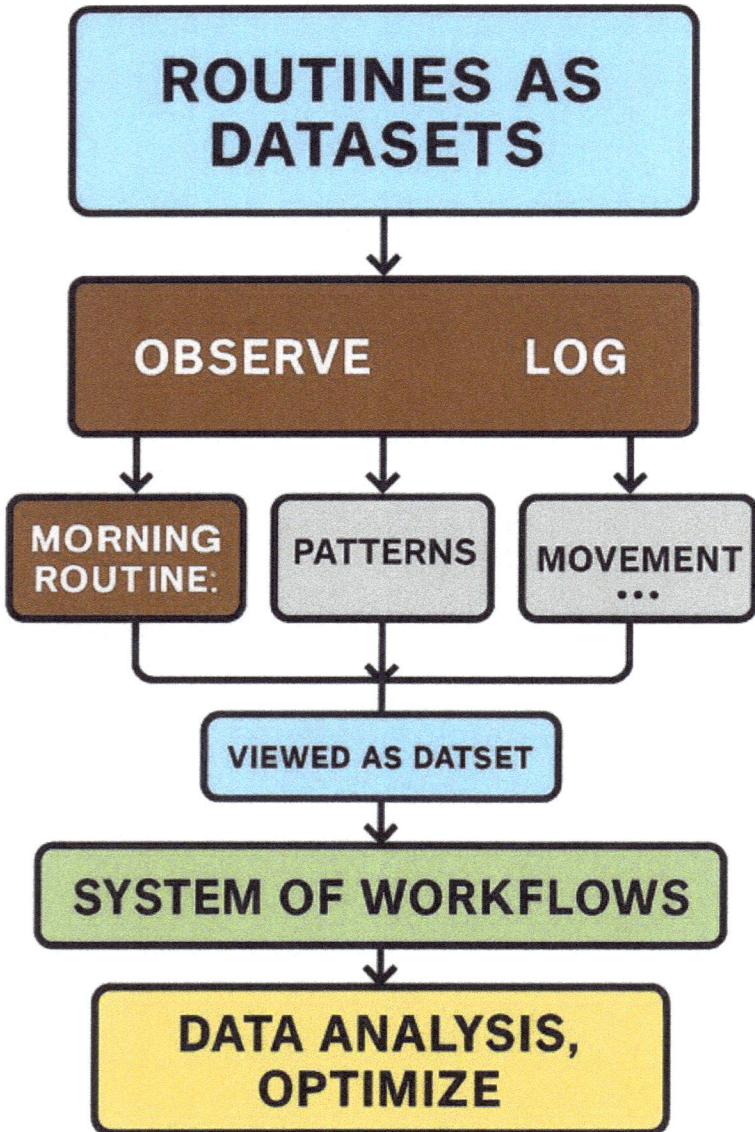

To start, consider logging your morning activities for a few days. Document what you do, when you do it, and how it makes you feel. This simple exercise will help you identify where your time is being spent and where disruptions occur. You might discover that checking your phone first thing in the morning consumes more time than you realize, which delays your preparation and increases your stress.

Think of your morning like a mini-process:

- Wake up
- Hygiene
- Breakfast
- Commute

If you're always rushing or stressed, something's off in the flow. Try tracking what's taking your time. Maybe a quick phone check turns into 20 minutes of scrolling.

In summary, the key to optimizing your mornings and overall routine lies in recognizing the data they provide. By observing, logging, and analyzing your actions, you can identify small changes that lead to significant improvements.

Meals and Preferences:

Across households worldwide, the daily act of preparing meals is a rich example of data-driven decision-

Morning Routine Tracker

Time	Activity	Mood	Energy
6:30 AM	Wake up	🙂	Low
7:00 AM	Shower & dress	🙂	Moderate
7:30 AM	Breakfast	😄	High
8:00 AM	Commute	😐	Moderate

making. Though often unrecognized as such, the process of meal planning and cooking involves a complex interplay of inventory management, preference analysis, and consumption forecasting. These culinary activities are not only about feeding ourselves but are also a demonstration of how data can be intuitively applied to enhance everyday life.

Each time you open the pantry or refrigerator to assess the available ingredients, you're engaging in a form of inventory analysis. This mental checklist allows you to determine what's on hand and what needs replenishing. Such assessments help avoid waste and ensure that resources are used efficiently. By tracking the usage rate of staples like rice or pasta, you can predict when you'll need to restock, thus preventing shortages and optimizing grocery shopping frequency.

Meal planning extends beyond just knowing what is available; it requires consideration of multiple variables such as dietary restrictions, family preferences, and nutritional needs. These factors form a dynamic data set that influences daily meal choices. For instance, if a family member is lactose intolerant, this information becomes a critical variable in planning meals that are inclusive and satisfying for everyone.

When deciding what to cook, you weigh factors such as the time available, the complexity of the recipe, and the anticipated response from those who will eat it. This decision-making process mirrors a predictive model where past experiences and current data inform future choices. For example, if a specific dish has been well-received in the past, it is likely to become a staple in your meal rotation. Moreover, the concept of demand prediction is vividly illustrated when preparing meals. Anticipating how much food will be needed based on previous consumption patterns helps in portion control and reduces food waste. If you know that your children tend to eat more after soccer practice, this insight enables you to prepare larger portions on those days.

Preferences, though often unspoken, are another layer of data that influences meal preparation. Over time, you build a mental database of likes and dislikes, which helps in making meals that are enjoyed by all.

In addition to planning and preparation, there is also a reactive component to meal management. Noticing that a particular ingredient is consistently left uneaten might prompt adjustments in purchasing habits or recipe selection. This feedback loop is a prime example of how data analysis can lead to continuous improvement in meal planning.

In essence, the kitchen is a microcosm of data analysis in action. By understanding and utilizing the data generated through meal preparation and consumption, individuals can create more efficient, enjoyable, and sustainable food practices. Recognizing these processes as data-driven doesn't just validate

DAILY DECISION PATTERNS

the complexity of everyday tasks but also empowers individuals to apply similar analytical thinking in broader contexts.

Daily Decision Patterns:

Every day, individuals make numerous decisions that, although seemingly trivial, are heavily influenced by data-driven insights. These decisions are often so ingrained in daily routines that their analytical nature goes unnoticed. The essence of data analysis lies in collecting information, recognizing patterns, and making informed decisions a process that is intuitively practiced by many in their everyday lives.

Consider the simple act of a grocery trip. Before heading to the store, one might mentally check what is already available at home, note which items are running low, and recall family preferences. This process is a form of inventory awareness, where past consumption data helps predict future needs. The decision to buy a smaller loaf of bread because the last one went moldy, or to purchase more yogurt during the summer because the kids consume more, is an example of predictive analytics at work. This form of analysis ensures that waste is minimized and shortages are avoided.

In addition to inventory checks, decisions made in the kitchen further illustrate the daily analysis of data. When preparing meals, factors such as available ingredients, family preferences, and time constraints are taken into account. This process is similar to creating a data model, where various variables are combined to produce an optimal outcome. Adjusting recipes based on taste tests, such as adding more salt or spices, is akin to a feedback loop where real-time data leads to immediate adjustments.

Furthermore, personal preferences play a significant role in daily decision-making. Although a formal list may not be kept, mental notes about who likes certain foods or who has dietary restrictions are maintained. This forms a preference database that influences meal selection, showcasing a type of filtering and ranking system in action. This mental database enables balancing multiple preferences to achieve satisfaction for all parties involved.

Visualization techniques also play a role in these daily routines. A sticky note on the fridge acts as a visual dashboard, while a calendar with meal plans serves as a time-series model. These tools help organize data, providing clarity and aiding in decision-making processes. Such visual aids are crucial for maintaining order and ensuring efficient execution of daily tasks.

In essence, the practice of data analysis is seamlessly integrated into daily life. Whether it is through meal planning or managing household inventories, individuals regularly engage in data-driven decision-making. By recognizing these patterns and becoming more aware of these practices, one can enhance efficiency, reduce waste, and improve overall satisfaction in daily routines. This awareness not only makes life easier but also fosters a more intentional approach to everyday tasks, basically leading to achieving better outcomes and a more organized lifestyle.

FEEDBACK LOOPS

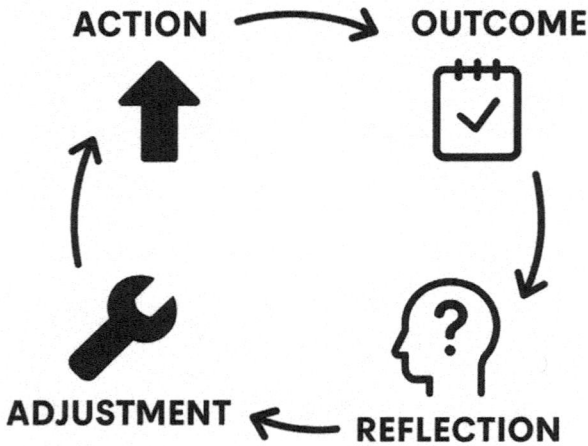

ACTION ⟶ **OUTCOME**

ADJUSTMENT ⟵ **REFLECTION**

Feedback Loops:

In data analysis, feedback loops serve as a fundamental mechanism for learning and adaptation. A feedback loop is a cyclical process that begins with an action, followed by an outcome, reflection, and adjustment. This iterative cycle allows individuals and systems to refine their behaviors and improve outcomes over time.

At its core, a feedback loop involves taking an action and observing the outcome. This observation is critical as it provides the data necessary for reflection. Reflection consists of evaluating the results of an action and determining whether it achieved the intended goals or outcomes. If the outcome is not satisfactory, the next step is to adjust the approach based on the insights gained from reflection and try again. This process repeats, allowing for continuous improvement.

Feedback loops can be observed in various aspects of daily life. Consider the example of fitness. When someone starts a new exercise routine, they might initially feel energized but notice fatigue setting in on subsequent days. By reflecting on this pattern, they may realize the need for more rest or nutritional adjustments, which can lead to changes in their routine. This is a closed feedback loop, where the cycle of action, reflection, and adjustment is completed, resulting in improved fitness outcomes.

A person might notice feeling sluggish after consuming fast food for lunch. Recognizing this pattern, they might opt for lighter meals, which can lead to improved focus and energy levels in the afternoon. The feedback loop here involves observing the impact of dietary choices and making informed adjustments to enhance well-being.

Social interactions are another area where feedback loops play a crucial role. For instance, if someone interrupts a colleague during a meeting and observes an adverse reaction, they might reflect on their communication style and decide to listen more actively in future interactions.

Spending Breakdown

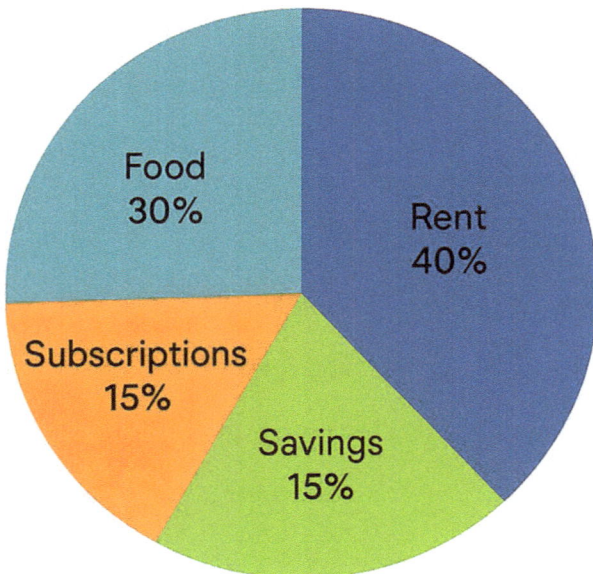

Food 30%
Rent 40%
Subscriptions 15%
Savings 15%

This reflection and adjustment process helps strengthen relationships and improve communication skills.

In family or team settings, recognizing patterns in behavior and outcomes can lead to improved routines and dynamics. For example, if a family notices that everyone is more relaxed and communicative during weekend dinners, they might make it a regular practice, enhancing family bonds.

The key to effective feedback loops is the willingness to observe and adapt. It requires an openness to change and a commitment to learning from experiences. By embracing feedback loops, individuals and organizations can cultivate a culture of continuous improvement, where data-driven insights inform better decisions and outcomes.

In conclusion, feedback loops are a powerful tool for personal and professional development. They transform experiences into opportunities for growth by turning reflections into actionable insights. Whether in health, social interactions, or professional settings, feedback loops help refine strategies, optimize performance, and achieve desired results. By understanding and leveraging feedback loops, we can navigate life's challenges more effectively, making informed decisions that enhance our overall quality of life.

4.   Smart Spending

Every purchase you make is part of a bigger financial story. Smart Spending is about taking charge of that story, using data and insight to make informed choices with your money. This chapter breaks it down into four key parts: Budgeting Basics, where you'll learn to plan and track spending; Shopping Strategies, to help you save while getting what you need; Understanding Consumer Data, so you know how your choices are influenced; and Financial Awareness, to build lasting habits for a healthier financial future. With the right mindset and a few practical tools, smart spending becomes second nature.

Budgeting Basics:

Budgeting is often perceived as a mundane task, relegated to the realms of spreadsheets and financial planning. However, in "Data Analysis for Everyone," budgeting is reimagined as a dynamic process of insight and discovery, reflecting not just numbers but the very priorities and habits that shape our lives. At its core, budgeting is a form of personal data analysis-a tool that can transform how we perceive and manage our financial resources.

The fundamental premise of budgeting is understanding the flow of money: what comes in and what goes out. By systematically tracking income and expenses, individuals create a personal dataset that offers a clear picture of their financial health. This process goes beyond merely listing expenses; it involves categorizing them into essentials, lifestyle choices, and irregular payments. Essentials might include rent and utilities, while lifestyle choices could cover dining out and entertainment. Irregular expenses, such as medical bills or car repairs, add another layer of complexity. This classification enables individuals to identify spending patterns, highlighting areas where adjustments can be made to align spending with personal values and goals.

One of the key insights from budgeting is identifying spending patterns. Over time, as data accumulates, specific trends become apparent. Perhaps there is a tendency for frequent impulse purchases, or maybe subscriptions are consuming a larger portion of the budget than initially realized. Seasonal spending spikes, such as those during holiday periods or school terms, also become evident. Recognizing these patterns allows for strategic budget adjustments, setting spending limits, and planning for future financial needs.

In today's digital age, the tools for tracking financial data have undergone significant evolution. While complex software exists, the essence of effective budgeting lies in consistency rather than complexity. Simple tools, such as spreadsheets or budgeting apps, suffice for most users. The key is regular review and analysis of financial data, which enhances the value of the insights gained.

Foreseeing future expenses is another crucial component of budgeting. By analyzing past spending patterns, individuals can predict upcoming financial commitments, such as monthly bills or annual insurance premiums. This foresight empowers individuals to plan for life changes, whether it be expanding a family, relocating, or retiring. Data-driven outlook transforms budgeting from a reactive to a proactive exercise.

Budgeting also involves making informed trade-offs. With finite resources, decisions must be made about where to allocate funds. Should extra money be spent on upgrading a car or saved for a family vacation? Is frequent dining out a sustainable long-term habit? Historical spending data provides clarity, enabling individuals to make these decisions based on evidence rather than speculation.

An adequate budget also includes provisions for unexpected expenses. Building an emergency fund and setting savings goals are integral parts of financial planning. These elements ensure that when unforeseen circumstances arise, there is a financial cushion to fall back on. This aspect of budgeting is not just about financial security but also about peace of mind.

Budgeting serves as an educational tool, particularly within families. Involving children in the budgeting process teaches them the value of money, the necessity of trade-offs, and the importance of planning for future needs. By engaging with budgeting from a young age, individuals develop a lifelong appreciation for financial literacy.

In essence, budgeting is not just about cutting costs; it is about gaining a comprehensive understanding of where money goes, why it goes there, and how to ensure it aligns with one's values and aspirations. By viewing finances through the lens of data, individuals gain confidence, control, and the ability to make choices that reflect their true priorities.

Shopping Strategies:

Every visit to a grocery store is a dynamic exercise in data analysis. As shoppers navigate aisles, they engage in a continuous process of decision-making driven by both real-time information and historical data. This interaction involves a multitude of strategies that, consciously or subconsciously, utilize data analysis principles to optimize purchases and manage budgets effectively.

One of the fundamental aspects of shopping involves being aware of inventory. Shoppers often mentally track their pantries' stock levels, recalling what items are already at home and what are running low. This informal process of inventory management is crucial in ensuring that purchases align with actual needs, thereby preventing both shortages and waste.

This mental checklist is a form of data management that balances current inventory against anticipated consumption.

As consumers evaluate products, they engage in real-time analysis, comparing prices, quality, and offers to make informed decisions. When faced with options like a cheaper brand versus an organic one or a buy-one-get-one deal, shoppers weigh these factors based on personal values, health goals, and budget constraints. This scenario exemplifies multi-criteria decision-making, a core aspect of data analysis that involves assessing multiple variables to make informed choices.

Predictive analytics also plays a significant role in shopping strategies. Consumers often rely on past experiences to inform current decisions. For instance, if a shopper recalls discarding moldy bread from a previous purchase, they might opt for a smaller loaf this time to avoid waste. Similarly, past consumption patterns, such as increased yogurt intake during summer, guide adjustments in purchase quantities. This predictive behavior helps in maintaining a balance between supply and demand, minimizing both excess and scarcity.

Budget management is another critical component of shopping. As shoppers fill their carts, they often keep a running total in their minds or use digital tools to track expenses.

Upon nearing a budget limit, they may make trade-offs, such as choosing store-brand items over premium brands or removing non-essential products from their cart. This real-time budgeting requires a constant balancing act between cost-effectiveness and meeting needs.

Retailers, too, engage in data analysis to influence consumer behavior. Loyalty cards and coupons serve as tools for personalized marketing strategies. By tracking purchase histories, stores can predict consumer interests and send targeted incentives to encourage specific buying behaviors. This form of consumer analytics not only enhances customer experience but also drives sales.

Store layouts and product placements are strategically designed using data insights. Essentials like milk and eggs are often placed far apart to increase browsing time and encourage impulse buys. Heatmaps and path analysis help stores understand customer movement patterns, enabling them to optimize layout designs and highlight profitable items, ultimately increasing overall sales.

Furthermore, technological advancements such as self-checkout systems provide real-time inventory updates, allowing stores to detect popular items, automate

Understanding Consumer Data

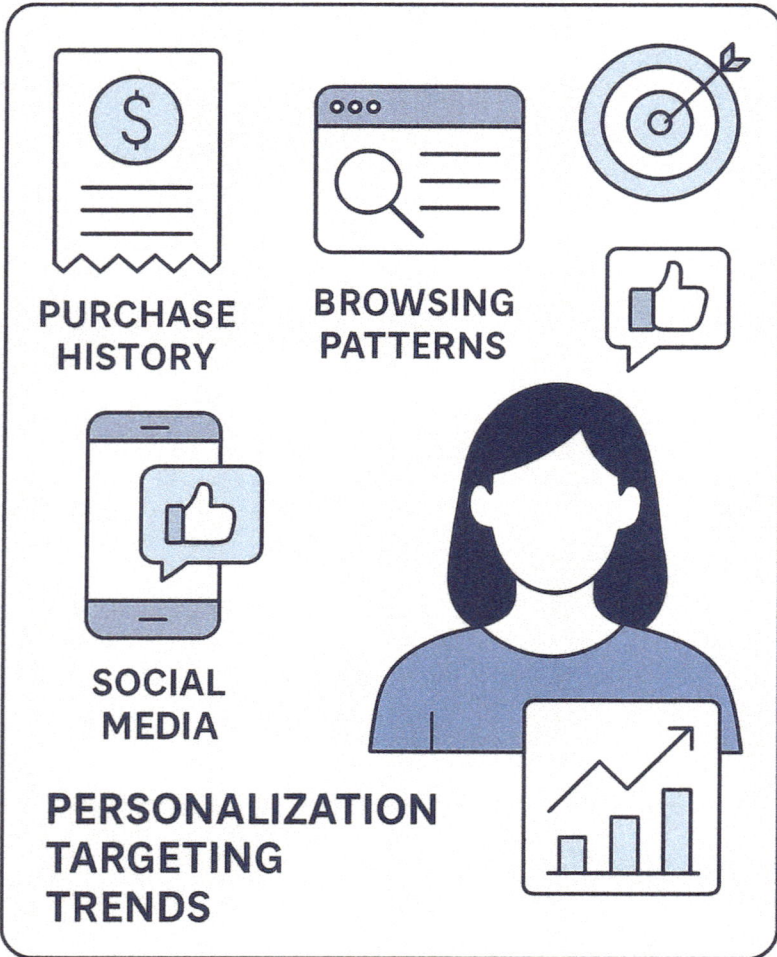

PURCHASE HISTORY

BROWSING PATTERNS

SOCIAL MEDIA

PERSONALIZATION
TARGETING
TRENDS

reorders, and monitor for potential theft or errors. This integration of technology in retail operations exemplifies how data not only transforms consumer decision-making but also enhances the operational efficiency of stores.

In essence, shopping is a shared data experience where both consumers and retailers apply data analysis to make informed decisions and influence outcomes. Recognizing this interplay empowers shoppers to exercise greater control and make more intentional, data-driven choices during their shopping journeys.

Understanding Consumer Data:

In today's digital age, the way businesses interact with consumers has undergone a dramatic transformation. With the proliferation of data, companies have a wealth of information at their fingertips, allowing them to understand and predict consumer behavior more accurately than ever before. This chapter explores the intricacies of consumer data, highlighting its significance and the methods for harnessing it effectively.

Consumer data encompasses a wide array of information collected from various sources, including purchase history, online browsing patterns, social media interactions, and other similar data. This data is not merely a collection of numbers and facts; it represents the behaviors, preferences, and needs of consumers.

By analyzing this data, businesses can gain insights into consumer preferences, enabling them to tailor their products, services, and marketing strategies to meet the specific needs of their audience.

One of the primary benefits of understanding consumer data is the ability to tailor marketing efforts to individual needs. With detailed consumer profiles, companies can create targeted marketing campaigns that resonate with specific customer segments. For instance, if data analysis reveals that a particular group of consumers frequently purchases eco-friendly products, a company might target this group with promotions for their sustainable product lines. Personalization not only increases the effectiveness of marketing campaigns but also enhances customer satisfaction and loyalty.

Also, consumer data analysis helps businesses optimize their inventory and product offerings. By tracking purchasing trends and consumer feedback, companies can identify which products are in high demand and which are underperforming. This information enables businesses to adjust their inventory levels and product offerings accordingly, thereby reducing waste and enhancing profitability. For example, a retailer might notice a spike in demand for a particular fashion item during a specific season and can adjust their stock levels to meet this demand.

Another critical application of consumer data is in predicting future trends. By analyzing historical data, businesses can forecast upcoming consumer behaviors and market trends. This predictive analysis enables companies to stay ahead of the competition by anticipating changes in consumer preferences and adjusting their strategies accordingly. For instance, a company might use data analysis to predict a growing interest in health-conscious food products. It can then expand its product line to include more health-oriented options.

However, with the vast amount of consumer data available, businesses face the challenge of managing and protecting this information. Data privacy and security are paramount, as consumers are increasingly concerned about how their personal information is used and protected. Companies must ensure compliance with data protection regulations and implement robust security measures to protect consumer data from breaches and misuse.

Understanding consumer data is crucial for businesses looking to thrive in a competitive market. By leveraging consumer data, companies can refine their marketing efforts, refine their product offerings, and anticipate future trends. However, it is equally important to address data privacy concerns and ensure the ethical use of consumer information.

As businesses continue to navigate the ever-evolving landscape of consumer data, those that effectively harness this powerful resource will be well-positioned for success.

Financial Awareness:

Financial awareness is a crucial skill that enables individuals to make informed decisions about their finances, in essence leading to economic stability and security. At its core, financial awareness involves understanding how cash flows in and out of your life, recognizing spending patterns, and making conscious choices that align with personal values and goals.

One of the foundational aspects of financial awareness is tracking your income and expenses. This process involves listing all sources of income and categorizing expenses to see where your money goes each month. By doing this, you create a personal financial dataset that reveals whether you are saving or overspending. It enables you to identify areas where you can cut back or reallocate resources to better align with your financial goals.

Understanding the distinction between needs and wants is another crucial aspect of financial awareness. Needs are essentials required for survival, such as food, shelter, and healthcare, while wants are non-essential items that provide comfort or pleasure. By distinguishing between these two, you can prioritize your spending on necessities while still allowing room for occasional indulgences without derailing your financial plans.

Budgeting plays a crucial role in enhancing financial awareness. A budget is a plan for your money that helps you allocate funds towards different categories such as savings, essentials, and entertainment. It serves as a guide to ensure that you do not exceed your income and helps you make adjustments as needed. Budgeting isn't merely about restricting spending; it's a tool for understanding your spending habits and making informed decisions about where your money should be allocated.

Future cost estimation is another crucial aspect of financial awareness. By analyzing past spending patterns, you can accurately predict upcoming expenses and prepare accordingly. This might include setting aside money for annual bills, such as insurance premiums or taxes, and planning for one-time expenses like vacations or home repairs. By anticipating these costs, you reduce the likelihood of financial surprises and maintain better control over your finances.

Saving is a vital practice within financial awareness. Building an emergency fund, for instance, is crucial for handling unexpected expenses without going into debt. Additionally, setting savings goals for specific purposes, such as education, retirement, or a significant purchase, helps you stay focused and motivated to achieve these objectives. Savings provide a safety net that enhances financial security and peace of mind.

Financial awareness also involves understanding the impact of debt on your financial health. Managing debt responsibly means understanding the terms of your loans, including interest rates and repayment schedules, and making consistent payments to avoid accumulating additional interest or penalties. Reducing high-interest debt should be a priority to free up more resources for savings and other financial goals.

Finally, financial awareness includes being informed about financial products and services. This involves understanding how various investment vehicles operate, recognizing the benefits and risks associated with them, and selecting options that align with your risk tolerance and financial objectives. Being knowledgeable about financial products empowers you to make more informed decisions and capitalize on growth opportunities.

In summary, financial awareness is about being proactive with your finances. It involves tracking income and expenses, budgeting, saving, managing debt, and making informed decisions about financial products and services.

By developing these skills, you enhance your ability to navigate the economic landscape effectively, ensuring a more secure and prosperous future.

5. Feedback Loops in Daily Life

Recognizing Patterns:

Our daily lives are filled with repetitive sequences that, often unnoticed, form the backbone of our routines. These sequences, or patterns, are not merely habits but repositories of valuable data that can be harnessed to optimize various aspects of life. Recognizing these patterns is a crucial step in data analysis, enabling us to predict outcomes, make informed decisions, and enhance our overall efficiency.

Action → Outcome → Reflection → Adjustment

It's how you naturally learn:

- **You eat junk → feel bad → eat lighter next time**
- **You stay up late → wake up tired → aim for better sleep**
- **You overspend → feel stress → cut back next week**

In addition to personal routines, patterns emerge in our interactions with others. Family and group dynamics often exhibit predictable behaviors, such as a household's tendency to order takeout on Fridays or a team's increased engagement during midweek meetings. Identifying these group patterns can lead to improved family routines, more effective meeting planning, and better anticipation of group needs. Behavioral feedback, such

RECOGNIZING DAILY PATTERNS

Waking up

Brushing teeth

Light energy dip

Snacking

Creative ideas during interview

Friday takeout

as recognizing that skipping breakfast leads to sluggishness, can inform personal rules that enhance daily functioning.

Recognizing patterns extends beyond mere observation-it involves envisioning and exploring possibilities. When patterns are identified, they enable short-term estimation, such as predicting traffic conditions or estimating sleep duration. This foresight enables proactive living, allowing individuals to plan meals based on their dietary patterns or develop study schedules that align with their energy rhythms. By using patterns consciously, individuals can set goals, reduce uncertainties, and enhance satisfaction.

The ability to recognize and trust patterns equips us with a powerful tool for personal and professional growth. It transforms routine observations into strategic insights, facilitating more informed decision-making and effective resource management. As we become more attuned to the rhythms of our lives, we gain greater control over our choices, fostering an environment where data-driven decisions become second nature. This chapter serves as a gateway to understanding how everyday patterns can be leveraged to improve our lives, setting the stage for more advanced data analysis techniques that follow.

Adjusting Routines:

Many of us operate on autopilot, following routines that, while familiar, may not always serve our best interests. Recognizing this, the key to enhancing our daily lives lies in adjusting these routines with a keen awareness of the data they generate. Our daily patterns, from morning rituals to evening wind-downs, are rich with data that can be analyzed and optimized for improved efficiency and satisfaction.

Understanding the nuances of our routines begins with observation. Each action in a routine is a data point that can reveal insights into our habits. By logging activities, noting the time they consume, and identifying points of friction, one can start to see patterns emerge. For instance, a morning routine might include checking a smartphone upon waking, which can lead to delays and added stress. Recognizing this as a pattern allows for adjustments, such as setting a phone-free period right after waking, which can significantly reduce morning chaos.

Once you see your patterns, you can:

- Schedule focused work when your energy peaks
- Batch chores during low-brain-power moments
- Pre-plan meals based on who eats what, and when

Even feelings are feedback. If you're constantly overwhelmed or drained, your life might be trying to tell you something.

Incorporating these insights into daily life doesn't require drastic changes; it simply involves making informed choices. Instead, minor, intentional adjustments, such as altering meal composition or adjusting sleep schedules, can lead to significant improvements. The process is iterative, with each change providing new data to guide further adjustments.

Ultimately, treating routines as datasets allows for a dynamic approach to personal improvement. By logging, observing, and adjusting, one can refine one's routines to better align with personal goals and preferences, leading to a more efficient and satisfying daily life. This method of routine management not only enhances individual productivity but also contributes to overall well-being, demonstrating the power of data in everyday life.

Learning from Outcomes:

When it comes to data analysis, understanding how to learn from outcomes is a pivotal skill. It transforms mere results into actionable insights, enabling individuals to make informed decisions based on past experiences. This process begins with the concept of feedback loops, which are fundamental to learning from outcomes. A feedback loop is a cyclical process involving four key steps: action, outcome, reflection, and adjustment.

When an action is taken, it leads to a specific outcome, which is then reflected upon, allowing for an evaluation of its success or areas for improvement. The reflection phase is crucial, as it involves a deep analysis of what worked, what didn't, and why. This introspection leads to adjustments in future actions, creating a continuous cycle of learning and improvement. Each completed loop enhances understanding and promotes growth, making it a powerful tool for personal and professional development.

Consider the example of a fitness routine. An individual may start a new exercise regimen and notice how it affects their energy levels. By reflecting on the outcome, whether they feel more energized or fatigued, they can adjust their routine to optimize results, such as altering the intensity or timing of workouts. This iterative process of trying, observing, and tweaking is the essence of learning from outcomes.

In a professional context, learning from outcomes involves analyzing data-driven results to refine strategies and operations. For instance, a team might launch a marketing campaign and track its effectiveness using metrics such as engagement rates and sales conversions. By examining these outcomes, they can identify which aspects of the campaign were successful and which need refinement.

This analysis enables better-informed decisions for future initiatives, thereby enhancing overall effectiveness and efficiency.

Learning from outcomes is not limited to quantitative data. Qualitative insights, such as customer feedback or team morale, also provide valuable information. For example, a company might conduct employee surveys to gauge satisfaction and identify areas for improvement. By acting on this feedback, they can foster a more positive work environment, which in turn can lead to increased productivity and employee retention.

The ability to learn from outcomes is a critical component of adaptive learning, a method that emphasizes flexibility and responsiveness to change. In rapidly evolving environments, the capacity to adapt based on outcomes is essential for sustained success. Adaptive learning, in this context, refers to the ability to adjust one's approach based on the results of past actions. This requires a mindset that views outcomes not as final judgments but as opportunities for growth and development.

Ultimately, learning from outcomes empowers individuals and organizations to transform data into meaningful insights. It encourages a culture of continuous improvement, where each outcome, whether positive or negative, is seen as a stepping stone toward greater understanding and capability. By cultivating this approach, one can navigate the complexities of modern life with greater confidence and clarity, constantly refining one's path based on the lessons learned from each experience.

Continuous Improvement:

Continuous improvement is a fundamental concept in data analysis, emphasizing the importance of iterative progress through minor, consistent enhancements. This approach is not only applicable in professional settings but also in personal growth and everyday decision-making. At its core, continuous improvement revolves around the feedback loop —a cyclical process that involves taking action, observing outcomes, reflecting on the results, and making adjustments accordingly.

The feedback loop is a crucial mechanism that enables ongoing refinement and optimization. It begins with an action, whether it's implementing a new strategy at work, trying a different study method, or altering a daily routine. Following the action, outcomes are observed and assessed. This stage is critical as it provides the data needed to evaluate effectiveness. Reflection then plays a pivotal role, where one analyzes the results, identifying what worked well and what could be improved. Finally, adjustments are made based on these insights, leading to a new cycle of action and observation.

In personal development, continuous improvement can manifest in various forms. For instance, if an individual aims to enhance their fitness level, they might start by tracking their physical activity and dietary habits. By regularly reviewing this data, they can identify patterns, such as which exercises yield the best results or how different foods impact energy levels. This reflective process enables them to make informed decisions, such as modifying their workout routines or adjusting their dietary choices, leading to improved health outcomes.

Similarly, in professional environments, continuous improvement is crucial for maintaining a competitive advantage and achieving long-term success. Businesses often employ methodologies like Kaizen or Six Sigma, which are rooted in the principles of continuous improvement. These frameworks encourage companies to foster a culture of constant evaluation and refinement, where every employee is empowered to contribute ideas for improvement. By systematically analyzing performance metrics and customer feedback, organizations can identify inefficiencies and develop strategies to enhance productivity and service quality.

Continuous improvement also extends to learning and skill development. In educational settings, students can benefit from adopting a mindset focused on iterative growth. By setting specific learning goals and regularly assessing their progress, students can tailor their study strategies to suit their needs better. This might involve experimenting with different note-taking techniques, adjusting study schedules, or incorporating new resources. Over time, these minor adjustments accumulate, leading to significant advancements in knowledge and skills.

The practice of continuous improvement is about embracing change and being open to experimentation. It requires a proactive attitude towards learning from both successes and failures. By viewing setbacks as opportunities

for growth rather than obstacles, individuals and organizations can foster resilience and adaptability. This mindset not only enhances personal and professional development but also promotes a culture of innovation and excellence.

In conclusion, continuous improvement is a dynamic process that drives progress through small, deliberate changes. By leveraging feedback loops and embracing a mindset of perpetual learning, individuals and organizations can achieve sustained growth and success. Whether applied to personal goals, professional endeavors, or educational pursuits, the principles of continuous improvement empower us to navigate challenges and seize opportunities with confidence and clarity.

6. Patterns,Forecasts,&Possibilities

Identifying Trends:

In data analysis, recognizing patterns is a crucial skill that enables us to make informed decisions based on historical data. Patterns are essentially recurring sequences or structures that provide insights into future occurrences or behaviors. By identifying these trends, we can predict outcomes, optimize processes, and adjust strategies to better align with our objectives.

Understanding the significance of trends requires us to examine data with a discerning eye, seeking out regularities that may not be immediately apparent. This involves exploring data sets over time to observe changes and consistencies. For example, a business might track sales data over several months or years to identify seasonal peaks and troughs, enabling it to adjust stock levels and marketing efforts accordingly.

In our daily lives, trend identification is evident in simple actions, such as tracking our sleep patterns or monitoring our spending habits. By doing so, we can identify patterns that help us improve our health or manage our finances more effectively. For instance, noticing that we feel more rested when we go to bed at a specific time can lead us to adjust our bedtime routine for better sleep quality.

The process of identifying trends often involves data visualization tools like graphs and charts, which make it easier to spot patterns at a glance. These visual aids are invaluable as they transform abstract numbers into a more

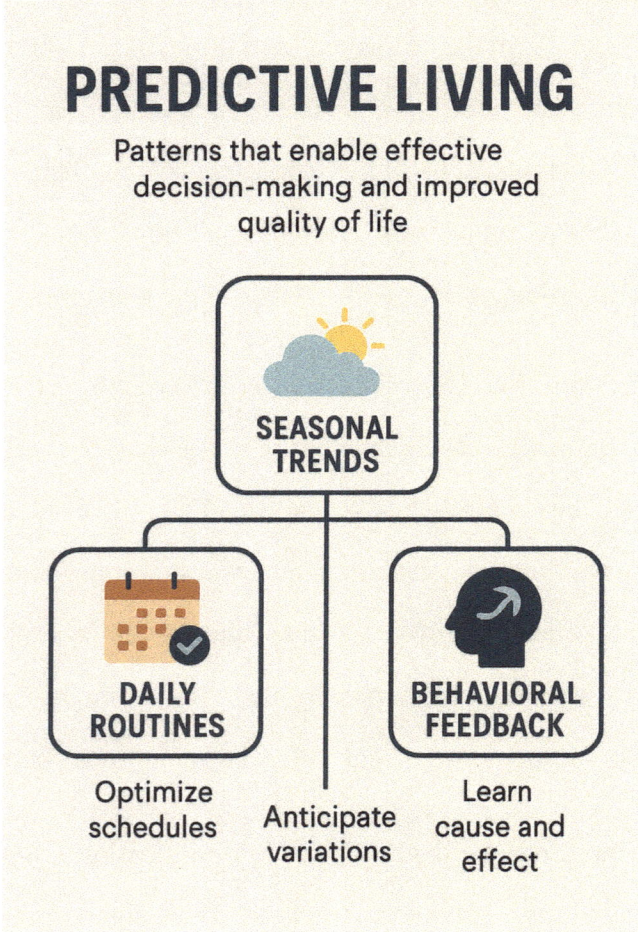

tangible form that can be quickly interpreted. For example, a line graph showing a steady increase in temperature over several decades can vividly illustrate climate change trends.

Additionally, the identification of trends is not just about looking at past data but also about predicting future developments. In the business world, trend analysis can guide strategic planning, helping companies anticipate market shifts and consumer behavior changes. It allows organizations to be proactive rather than reactive, providing a competitive edge in rapidly evolving markets.

However, identifying trends comes with its challenges. It requires a balance between recognizing true patterns and avoiding overfitting, where one sees patterns that are not statistically significant. This is where statistical tools and techniques come into play, offering methods to validate the trends we perceive. Techniques like regression analysis, moving averages, and other statistical tests help ensure that the trends identified are reliable and not merely coincidental.

In addition, the context in which data exists is crucial for accurate trend identification. Understanding external factors that might influence data trends, such as economic conditions, technological advancements, or socio-political changes, is essential. These factors can significantly impact the data and, consequently, the trends we observe.

Ultimately, the ability to identify trends is a powerful tool in the data analyst's toolkit, enabling more effective decision-making and strategic planning. It transforms raw data into actionable insights, fostering a deeper understanding of the environments in which we operate. By mastering the art of trend analysis, we not only enhance our analytical capabilities but also our ability to predict and influence future events.

Predictive Living:

Our daily lives are filled with patterns that, when recognized, can lead to more effective decision-making and improved quality of life. These patterns manifest in various forms, such as daily routines, seasonal trends, and behavioral feedback, all of which contribute to the concept of predictive living. By understanding these patterns, individuals can optimize their schedules, enhance their health, and reduce stress.

Daily routines are more than habits; they are time-based data series that provide insights into our productivity and well-being. For instance, people often discover they are more productive in the morning and experience a dip in energy around mid-afternoon. Acknowledging these natural rhythms allows individuals to plan tasks according to their peak energy levels, thereby optimizing their daily schedules. Furthermore, recognizing when to focus, rest, or socialize based on these routines can lead to more efficient time management.

Seasonal trends also play a significant role in predictive

living. Patterns such as higher electricity bills during winter months, the onset of cold symptoms during seasonal changes, and increased outdoor activities in spring and summer, all provide valuable insights. By anticipating these seasonal variations, individuals can make informed decisions regarding budgeting, clothing purchases, and vacation planning. The ability to predict these trends enables better preparation and adaptation to changing circumstances.

Behavioral feedback is another crucial component of predictive living. It involves observing the cause-and-effect relationships in one's behavior, such as feeling sluggish after skipping breakfast or experiencing better sleep quality after reading before bed. Over time, these observations become personal "rules" that guide behavior, akin to observational learning. By adhering to these self-imposed guidelines, individuals can enhance their overall well-being and productivity.

Patterns do not only exist within individuals; they also emerge within families, workplaces, and communities. Recognizing group trends, such as children becoming cranky at the same time each day or a team being more engaged during midweek meetings, can lead to improved family routines and more effective meeting planning. By anticipating group needs, individuals can foster a more harmonious and productive environment.

Once these patterns are identified, the next step is to make predictions based on them. Short-term anticipating becomes possible, allowing individuals to anticipate outcomes, such as the amount of sleep they will get if they go to bed at a specific time or the cost of a trip if gas prices continue to rise. These forecasts, often based on personal data, empower individuals to make proactive decisions.

Prediction extends beyond prevention; it opens the door to proactive living. By planning family meals based on dietary patterns, building study plans around energy rhythms, or choosing investments based on past spending trends, individuals can set goals, reduce surprises, and increase satisfaction. Patterns, when used consciously, provide insights that can lead to a more intentional and fulfilling life.

Predictive living involves paying attention to the rhythms of one's day, week, and year to gain control over choices. By recognizing and trusting these patterns, individuals can anticipate needs, prepare effective responses, and explore potential possibilities. At its core, data analysis is about identifying patterns and asking, "What's likely to happen next?" This mindset not only enhances personal decision-making but also fosters a deeper understanding of the world around us.

Exploring Future Scenarios:

In data analysis world, exploring future scenarios is a crucial exercise that allows individuals and organizations to prepare for potential outcomes. This practice involves predicting and planning for various possibilities based on current and historical data. By examining patterns and trends, one can anticipate what might happen under different circumstances, thereby reducing uncertainty and enhancing decision-making.

The process begins with gathering relevant data, which forms the foundation for any predictive analysis. This data is collected from diverse sources, including past performance metrics, market trends, and consumer behavior. Once collected, the data is cleaned and organized to ensure accuracy and relevance. This step is vital as it sets the stage for developing reliable forecasts.

With the data in hand, analysts employ various models and techniques to simulate different scenarios. These models range from simple trend analysis to complex machine learning algorithms that can handle large datasets and multiple variables. The choice of model depends on the nature of the data and the specific questions being addressed. For instance, a simple linear regression may suffice for a straightforward trend projection, while more sophisticated models, such as neural networks, may be needed for complex, non-linear relationships.

Scenario planning involves not just predicting the most likely outcome but also considering alternative futures. This means creating a range of possible scenarios, each representing different assumptions about key variables. For example, a business might explore scenarios where market demand either increases or

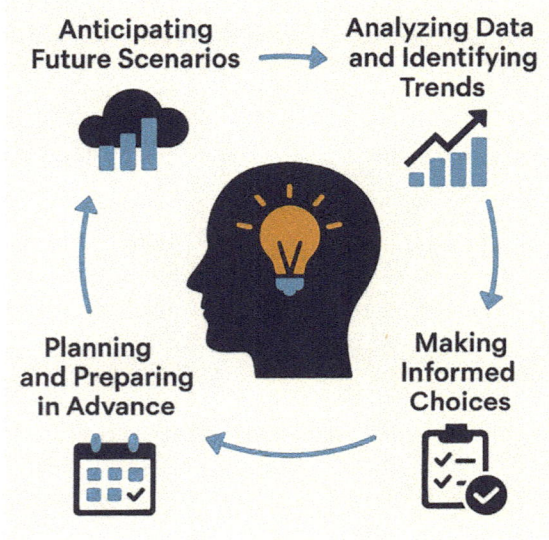

decreases significantly, or where new regulations impact operations. By considering these alternatives, organizations can develop robust strategies that are adaptable across various potential futures.

An essential part of exploring future scenarios is the use of visualization tools to effectively communicate findings. Visualizations such as graphs, charts, and dashboards help convey complex data insights in an accessible way, making it easier for decision-makers to grasp potential implications and make informed choices. These tools are handy in highlighting trends, comparing different scenarios, and illustrating the potential impact of various strategic options.

Feedback loops play a crucial role in refining scenarios over time. As new data becomes available, it is essential to revisit and update predictions to ensure they remain relevant and accurate. This iterative process allows for continuous improvement and adaptation, ensuring that scenario planning remains a dynamic and responsive exercise.

Exploring future scenarios is not limited to the business world; it is equally applicable in personal contexts. Individuals can utilize these techniques to plan their personal finances, career paths, or health goals. By anticipating potential changes and preparing accordingly, people can make proactive decisions that align with their long-term objectives.

In essence, exploring future scenarios is about embracing uncertainty and transforming it into an opportunity for strategic foresight. It equips individuals and organizations with the tools to navigate complex environments, anticipate challenges, and seize emerging opportunities. By systematically analyzing data and envisioning multiple futures, decision-makers can enhance resilience and adaptability, ensuring that they are better prepared for whatever the future may hold.

Proactive Decision Making

In the realm of decision making, proactive strategies involve anticipating future scenarios and preparing for them in advance. This approach contrasts with reactive decision-making, where choices are made in response to events as they occur. Proactive decision-making is rooted in the ability to foresee potential outcomes and plan accordingly, thereby enabling individuals and organizations to navigate challenges more effectively.

The essence of proactive decision-making lies in its foundation of data analysis. By examining historical data and identifying patterns, decision-makers can predict future trends and prepare accordingly. This involves collecting relevant data, analyzing it for insights, and using these insights to inform future decisions. The process is iterative, with each decision providing new data to refine future predictions.

One of the key components of proactive decision making is scenario planning. This involves envisioning various potential futures and considering the implications of each. By doing so, decision-makers can develop strategies that are robust across multiple possible outcomes. Scenario planning fosters flexibility and adaptability, enabling swift adjustments when circumstances shift.

Another essential element is the use of prediction models. These models leverage statistical tools and algorithms to forecast future events based on existing data. By understanding the probability of different outcomes, decision-makers can prioritize resources and efforts toward the most likely scenarios. This predictive approach reduces uncertainty and enhances the ability to make informed choices.

Feedback loops play a crucial role in informed decision-making. By continuously monitoring the outcomes of decisions and comparing them to predictions, decision-makers can adjust their strategies in real-time. This ongoing process of feedback and adjustment ensures that decisions remain relevant and practical as new data becomes available.

Proactive decision making involves risk management. It requires identifying potential risks and developing mitigation strategies in advance. By anticipating challenges, decision makers can reduce the likelihood of negative outcomes and be better prepared to handle them if they occur. This risk-aware approach is integral to maintaining stability and achieving long-term goals.

Collaboration and communication are also vital in proactive decision-making. Engaging stakeholders in the decision-making process ensures that diverse perspectives are considered, leading to more comprehensive strategies. Open communication facilitates the sharing of information and ideas, fostering a culture of innovation and continuous improvement.

In practice, proactive decision-making can be applied in various contexts, ranging from business strategy to personal life choices. For instance, a company might use proactive decision-making to anticipate market trends and adjust its product offerings accordingly. Similarly, an individual might apply these principles to career planning, setting goals based on future industry developments.

Overall, proactive decision-making empowers individuals and organizations to shape their futures rather than merely react to them. By leveraging data analysis, scenario planning, predictive models, and feedback loops, decision-makers can make informed, strategic, and resilient choices. This forward-thinking approach not only enhances decision quality but also builds the capacity to thrive in an increasingly complex and changing world.

7.   Body Data

Health Metrics:

In our everyday lives, we are constantly surrounded by data that speaks to our health. This data comes from a myriad of sources, including our physical activity, dietary habits, and sleep patterns. By understanding and interpreting these health metrics, we can improve our overall well-being and make informed decisions that affect our daily lives.

The first primary health metric to consider is physical activity. This encompasses everything from the number of steps we take each day to the intensity of our workouts. Regular physical activity is crucial for maintaining cardiovascular health, building muscle, and enhancing mental well-being. Tools like pedometers or fitness apps on smartphones provide an accessible way to track this data. By analyzing patterns in our activity levels, we can set realistic goals, identify when we are most active, and recognize periods of inactivity that may require adjustment.

Diet is another critical component of health metrics. It involves not only what we eat but also when and how much we consume. Keeping a food diary or using nutritional apps can help in tracking calorie intake, identifying nutrient deficiencies, and observing how different foods affect our energy levels and mood. This data helps in making dietary adjustments that can lead to better health outcomes, such as increased energy levels and improved mental clarity.

Sleep is a fundamental health metric that significantly impacts our daily functioning and long-term health. It is essential to not only monitor the quantity of sleep but also its quality. Factors such as sleep duration, bedtime consistency, and interruptions during the night all contribute to the overall quality of sleep. Sleep tracking devices and apps can provide insights into sleep patterns, helping individuals to make necessary adjustments to improve rest, such as altering bedtime routines or adjusting sleep environments to reduce disturbances.

These health metrics are interconnected. For instance, poor dietary choices can lead to disrupted sleep, while insufficient sleep can affect physical activity levels and food choices the following day. Understanding these interconnections allows for a more holistic approach to health management. By analyzing these metrics together, individuals can identify patterns that may not be apparent when viewed in isolation.

The use of technology plays a significant role in collecting and analyzing health metrics. From wearable fitness trackers to smartphone applications, technology provides the tools needed to gather accurate data. These tools not only help track and analyze data but also offer personalized feedback and suggestions for improvement. However, it is essential to remember that technology should serve as a guide rather than a strict rulebook. Personal intuition and understanding of one's body should also play a role in interpreting health data.

The goal of monitoring health metrics is to foster a proactive approach to health management. By being aware of and analyzing these daily metrics, individuals can make informed decisions that promote long-term health and wellness. This proactive approach empowers individuals to take control of their health, leading to improved outcomes and a higher quality of life.

Sleep Patterns:

Sleep is a fundamental aspect of human life, intricately linked to our overall well-being and daily functioning. Understanding sleep patterns can offer valuable insights into optimizing rest and enhancing productivity throughout the day. In the context of data analysis, sleep becomes more than just a nightly routine; it is a rich source of information that can be systematically observed and improved.

One of the primary aspects of sleep patterns is the duration of sleep. While the traditional recommendation is around eight hours per night, individual needs can vary significantly. Tracking sleep duration over time can reveal personal patterns, such as whether you consistently get less sleep on weekdays compared to weekends. This data can prompt adjustments, such as prioritizing earlier bedtimes or creating more conducive sleep environments.

Consistency in sleep timing is another critical factor. Regular sleep and wake times support the body's natural circadian rhythms, which regulate sleepiness and alertness. By maintaining a consistent schedule, even on weekends, individuals can improve their sleep quality and daytime energy levels. This consistency acts as a stabilizing force, allowing the body to predict and prepare for sleep more effectively.

The quality of sleep is as important as its duration. Disruptions during the night, such as waking up frequently or having difficulty falling back asleep, can undermine the restorative benefits of sleep. Tracking these disruptions can help identify patterns or triggers, such as stress, caffeine intake, or screen time before bed. Addressing these factors can significantly enhance sleep quality, leading to better physical and mental health.

The concept of sleep stages offers another layer of insight. Sleep is not a monolithic state but comprises cycles of different stages, including light sleep, deep sleep, and REM (rapid eye movement) sleep. Each stage plays a distinct role in physical recovery, memory consolidation, and emotional regulation. By understanding and optimizing these stages, individuals can ensure they are getting the most out of their sleep.

Data-driven strategies for improving sleep patterns often involve simple yet effective changes. These can include setting a consistent bedtime routine, reducing exposure to screens before bed, and creating a comfortable sleep environment. The use of sleep tracking devices or apps can also provide valuable feedback, helping individuals identify areas for improvement and track progress over time.

Viewing sleep through the lens of data analysis empowers individuals to make informed decisions about their health. By identifying patterns and experimenting with changes, individuals can transform their sleep habits, resulting in improved mood, enhanced cognitive function, and greater overall well-being. The exploration of sleep patterns thus becomes a journey of personal discovery, revealing the profound impact that a good night's sleep can have on everyday life.

Fitness Tracking:

Fitness tracking is a process that involves collecting and analyzing data related to physical activity and health metrics. This practice is gaining popularity due to the rise of wearable technology and mobile applications designed to track various aspects of personal fitness. By leveraging these tools, individuals can gain deeper insights into their physical activity patterns, leading to more informed decisions about their health and wellness routines.

One of the primary benefits of fitness tracking is its ability to provide real-time data on physical activity. Wearable devices, such as smartwatches and fitness bands, can track steps taken, distance traveled, heart rate, and calories burned. These devices often synchronize with mobile apps, offering users a comprehensive view of their daily activities. This data can help highlight trends and patterns in physical activity, enabling users to set realistic goals and track their progress over time.

Another critical aspect of fitness tracking is goal setting. By understanding their current fitness levels, individuals can set achievable targets, such as increasing daily step counts or improving cardiovascular endurance. Fitness apps often include features that allow users to set these goals and receive notifications or reminders to stay on track. This proactive approach encourages consistency and motivation, which are essential for achieving long-term health goals.

Fitness tracking also facilitates the identification of gaps in physical activity. Users can review their data to determine when they are most and least active, enabling them to make necessary adjustments to their routines. For example, someone who notices a decline in activity on weekends might decide to incorporate a scheduled workout or leisure activity to maintain consistency.

Also, fitness tracking can be instrumental in monitoring specific health metrics, such as heart rate variability or sleep patterns. Many advanced fitness trackers can measure these metrics, offering insights into the user's overall health and well-being. For instance, tracking sleep can reveal patterns that affect daily energy levels and performance, prompting individuals to make lifestyle changes that enhance their rest quality.

In addition to personal benefits, fitness tracking can contribute to broader health research and public health initiatives. Aggregated data from fitness trackers can help researchers understand general trends in physical activity and health behaviors across different populations. This information can inform public health policies and programs designed to improve community health outcomes.

However, the use of fitness tracking is not without challenges. Privacy concerns regarding the collection and use of personal health data are significant. Users must ensure that their data is protected and that they understand how it is used by the companies that provide these services. Additionally, the accuracy of fitness trackers can vary, and users should be aware of potential discrepancies in data collection.

In conclusion, fitness tracking offers numerous advantages for individuals seeking to improve their health and wellness. By providing detailed insights into physical activity and health metrics, these tools empower users to make data-driven decisions

Tracking Time
Log daily activities and identify patterns

Focus Zones
Align tasks with peak energy periods

Pareto Principle
80%
Prioritize tasks with the greatest impact

Structured Routines
Batch similar tasks; reduce decision fatigue

about their fitness routines. Despite challenges such as privacy and data accuracy, the benefits of fitness tracking can significantly enhance an individual's ability to lead a healthier lifestyle.

Holistic Wellbeing:

The concept of holistic wellbeing encompasses a comprehensive approach to health that integrates physical, mental, and emotional aspects, recognizing that these dimensions are interconnected and impact each other. This perspective emphasizes the importance of balance and harmony in life, advocating for a lifestyle that nurtures all facets of human existence.

Physical health is often the most visible aspect of wellbeing and serves as a foundation for holistic health. It involves regular physical activity, balanced nutrition, adequate sleep, and preventative healthcare. Engaging in consistent exercise not only strengthens the body but also boosts mental health by reducing anxiety and depression. Nutrition plays a crucial role in physical and psychological wellness, as a balanced diet provides the necessary nutrients to support bodily functions and cognitive processes. Sleep, a frequently overlooked component of health, is vital for recovery and cognitive performance, influencing mood and decision-making abilities.

Mental health, another crucial pillar of holistic wellbeing, involves the development of cognitive skills, emotional resilience, and psychological flexibility. It requires conscious efforts to manage stress, cultivate positive relationships, and engage in activities that foster personal growth and satisfaction. Practices such as mindfulness and meditation can enhance mental health by promoting relaxation and awareness, helping individuals to focus on the present moment and reduce stress.

Emotional wellbeing, closely linked to mental health, involves understanding and managing emotions effectively. It requires the ability to express feelings appropriately, cope with life's challenges, and build satisfying relationships. Emotional intelligence, which encompasses self-awareness, empathy, and social skills, is crucial in maintaining emotional well-being. Developing emotional intelligence enables individuals to navigate social complexities, foster stronger relationships, and enhance personal and professional satisfaction.

The integration of these elements into a cohesive approach to health is the essence of holistic wellbeing. It encourages individuals to view health as a dynamic process rather than a static state. This approach advocates for the continuous pursuit of balance and harmony in all aspects of life, recognizing that well-being is a dynamic and evolving journey.

Holistic wellbeing also emphasizes the importance of environmental and social factors. A supportive community and a healthy environment are integral to achieving overall health. Social connections offer emotional support, alleviate feelings of isolation, and foster a sense of belonging. Similarly, a healthy environment, free from pollution and conducive to physical activity, supports physical and mental health.

Incorporating holistic practices into daily life requires intentionality and commitment. It involves setting realistic goals, monitoring progress, and making adjustments as needed. Tools such as wellness journals, fitness trackers, and nutritional apps can aid in this process by providing feedback and helping individuals stay accountable to their health goals.

Ultimately, holistic wellbeing is about empowering individuals to take charge of their health by making informed choices that align with their values and life goals. It encourages a proactive approach to health, where individuals actively engage in practices that foster a balanced and fulfilling life. By embracing a holistic perspective, individuals can enhance their quality of life and achieve a greater sense of well-being.

8.   Time and Productivity Analysis

Time Management Techniques:

Time is an invaluable resource in data analysis. Effective time management is not about cramming more tasks into an already packed schedule but rather about strategically aligning one's time with priorities and energy levels. This alignment fosters not just productivity but also a sense of control and accomplishment.

The first step in mastering time management involves tracking how time is currently spent. This involves maintaining a detailed log over several days to capture a representative snapshot of daily activities. Activities should be categorized into blocks such as work, rest, meals, errands, and screen time. This process reveals patterns and highlights areas where time might be wasted or underutilized. Identifying these patterns is akin to conducting a time audit, which serves as a foundation for making informed adjustments.

Once a clear picture of time allocation emerges, the next step is to identify "focus zones" and "time drains." Focus zones are periods when energy and concentration peak, allowing for more complex and demanding tasks to be tackled efficiently. Conversely, time drains are activities or periods that result in little productivity, often consuming more time than necessary. Recognizing these can help in scheduling tasks that require high concentration during focus zones and relegating less critical tasks to periods when energy naturally dips.

A key technique for improving time management is the application of the Pareto Principle, also known as the 80/20 rule. This principle suggests that 80% of outcomes often result from 20% of efforts. Identifying and prioritizing the 20% of tasks that contribute most significantly to desired outcomes can drastically enhance productivity. This involves critically evaluating tasks to determine which ones are paramount and focusing energy on these high-impact activities.

Additionally, creating structured routines can significantly reduce decision fatigue and streamline daily processes. By batching similar tasks and establishing dedicated time blocks for specific activities, individuals can minimize the cognitive load associated with frequent task switching. This structured approach not only enhances efficiency but also provides a predictable framework that can accommodate unexpected disruptions with minimal stress.

Incorporating digital tools can further augment time management efforts. Applications such as calendars, task boards, and time-tracking software provide a visual representation of commitments and progress. These tools facilitate better planning and allow for real-time adjustments based on daily fluctuations in energy and priorities.

Setting boundaries is another crucial aspect of effective time management. This involves learning to say no to tasks and commitments that do not align with one's primary goals. Reviewing time data regularly can empower individuals to make informed decisions about what to eliminate, delegate, or defer.

Finally, regular reflection is vital to maintaining effective time management. Just as data analysis involves continuous evaluation and adjustment, time management requires periodic reviews to assess what strategies are working and which need refinement. This reflective practice supports sustained improvement and ensures that time is consistently aligned with personal and professional values and goals.

By approaching time as a finite resource to be analyzed and optimized, individuals can transform their daily schedules into powerful tools for achieving both personal and professional objectives. This strategic alignment of time management techniques with one's goals and energy levels can lead to enhanced productivity, reduced stress, and a more balanced life.

Productivity Patterns:

In the realm of personal productivity, understanding patterns is akin to unlocking a treasure trove of insights that can significantly enhance both efficiency and effectiveness. Productivity is often misconstrued as merely being busy. Yet, actual productivity involves strategically aligning tasks with optimal energy levels, minimizing time-wasting activities, and fostering an environment conducive to focused work.

One of the foundational steps in identifying productivity patterns is time tracking. This involves logging daily activities, breaking them into categories such as work, leisure, and chores, and analyzing these logs to uncover where time is being invested. This process helps identify which activities consume more time than necessary and which ones contribute positively to productivity. For instance, you might discover that a significant portion of your day is spent on tasks that offer little return in terms of progress or satisfaction. By categorizing activities, you can pinpoint those that are genuinely beneficial and those that can be minimized or delegated.

The concept of energy management is equally critical in understanding productivity patterns. Not all hours of the day are created equal; some periods are naturally more suited for high-focus tasks. By identifying when you are most alert and energetic, you can schedule demanding tasks during these peak periods, ensuring that your best efforts are applied when your energy is at its highest. Conversely, routine or less demanding tasks can be scheduled during energy dips, allowing for optimal use of your time and energy.

Moreover, the 80/20 rule, also known as the Pareto Principle, plays a pivotal role in productivity patterns. This principle suggests that 80% of results often come from 20% of efforts. Identifying which tasks or activities constitute this crucial 20% can lead to significant improvements in productivity. By focusing on these high-impact activities, you can maximize outcomes with less effort, making your productivity efforts more efficient and rewarding.

Recognizing and mitigating common time traps is another essential aspect. These traps can include frequent email checking, multitasking, and back-to-back meetings without breaks. Each of these activities can drain time and energy, leading to reduced productivity. Implementing strategies such as batching similar tasks, setting specific times for checking emails, and ensuring breaks between meetings can help avoid these pitfalls.

Feedback loops are also integral to refining productivity patterns. Regularly reviewing your productivity logs can highlight patterns and areas for improvement. By asking questions like, "When was I most productive? What distracted me? What changes can I make for next week?" you create a cycle of continuous improvement. This iterative process enables adjustments based on past performance, resulting in progressively improved productivity outcomes.

In conclusion, understanding and optimizing productivity patterns involves a combination of strategic time tracking, energy management, prioritizing high-impact tasks, avoiding common time traps, and utilizing feedback loops for ongoing improvement. By approaching productivity with a data-driven mindset, individuals can transform their daily routines into efficient workflows that enhance both personal and professional success.

Balancing Work and Life:

In today's fast-paced world, finding harmony between professional responsibilities and personal life is a challenge many face. The key to achieving this equilibrium lies in understanding and applying data analysis principles to our daily routines. This approach enables individuals to make informed decisions, optimize their time, and improve overall well-being.

A pivotal step in balancing work and life is identifying where time and energy are spent. By logging activities throughout the day, individuals can create a dataset that highlights patterns in their daily routines. This data helps pinpoint areas of inefficiency, such as frequent social media breaks or prolonged meetings with minimal outcomes. Recognizing these patterns is the first step toward making necessary adjustments that promote a more balanced life.

Energy management is as crucial as time management. Understanding when one feels most alert and when distractions are prevalent can significantly impact productivity. For instance, creative tasks are often best tackled in the morning when energy levels are high, while routine tasks might be more suited for the afternoon. This alignment of tasks with energy levels ensures that time is used more effectively, reducing stress and increasing output.

Setting priorities is another essential aspect of achieving balance. The Eisenhower Matrix, a tool for sorting tasks by urgency and importance, can help individuals focus on meaningful work rather than getting bogged down by busywork. By categorizing tasks into urgent and important, important but not urgent, urgent but not important, and neither urgent nor important, individuals can streamline their workflow and devote more time to personal pursuits.

Creating structured routines reduces decision fatigue, which often leads to burnout. Establishing systems such as batching similar tasks, using time blocks for focused work sessions, and allowing buffer time for unexpected events can make daily life more predictable and manageable. This predictability is key to maintaining a steady work-life balance.

Digital tools also play a significant role in balancing work and life. Applications like Google Calendar, Trello, or Notion help visualize daily tasks and commitments, providing structure and clarity. Time tracking tools like Toggl or Clockify can offer insights into how time is spent, enabling individuals to make data-driven decisions about where to cut back or invest more effort.

Boundary setting is crucial for maintaining a healthy work-life balance. By analyzing time data, individuals can decide what activities to eliminate, delegate, or decline. This ability to say no is essential for preserving time for activities that align with personal values and goals.

Regular reflection is vital for continuous improvement. Taking time each week to review what worked, what was stressful, and what needs change fosters long-term awareness and enhancement of work-life balance. This reflective practice ensures that adjustments are made proactively, rather than reactively, leading to a more intentional and fulfilling life.

By applying data analysis principles to the management of time and energy, individuals can achieve a harmonious balance between work and personal life. This approach not only enhances productivity but also contributes to greater satisfaction and well-being.

Maximizing Efficiency:

Efficiency in data analysis is about optimizing processes and utilizing resources effectively. By harnessing data-driven strategies, individuals and organizations can streamline operations, reduce waste, and enhance productivity. The first step towards achieving efficiency is understanding where time and resources are currently being spent. This requires a detailed logging of activities and tasks over a period, which serves as a baseline dataset for analysis.

Once this data is collected, identifying patterns and time-wasting activities becomes crucial. Common inefficiencies include excessive social media breaks, prolonged meetings with minimal outcomes, and lengthy commutes that sap energy. Recognizing these patterns is akin to analyzing a financial budget where wasteful expenditures are identified and curtailed.

A key aspect of maximizing efficiency is aligning tasks with energy levels. Productivity is not solely about time management but also about energy management. Observing when one feels most alert or prone to distractions can help in scheduling tasks that match these energy windows. For instance, creative tasks might be best tackled in the morning when energy levels are high, while routine administrative work might be more suitable for the afternoon.

Prioritization is another critical component of efficiency. Tools like the Eisenhower Matrix can help categorize tasks based on urgency and importance, allowing individuals to focus on meaningful progress rather than busywork. This method helps in reducing functions that are neither urgent nor important, thereby freeing up time for critical activities.

Creating systems and routines is essential for reducing decision fatigue and making productivity predictable. Batching similar tasks, using time blocks for focused sessions, and setting aside buffer time for unexpected events can help maintain a steady workflow. These systems ensure that productivity is sustainable over the long term.

Digital tools play a significant role in maximizing efficiency. Applications like Google Calendar, Trello, or Notion for task management, and time-tracking tools like Toggl or Clockify provide structure and visibility into daily activities. These tools transform a chaotic day into a visual workflow, facilitating better planning and execution.

Setting boundaries is equally vital in achieving efficiency. By reviewing time data, one can determine which activities to eliminate, delegate, or decline to ensure alignment with core goals. This data-driven boundary setting empowers individuals to say no to distractions and yes to what truly matters.

Ultimately, regular reflection is crucial for maintaining and enhancing efficiency. Just as one might review a budget or fitness log, spending a few minutes each week to assess what worked, what was stressful, and what needs to change can lead to long-term improvements. This feedback loop of tracking, analyzing, and adjusting helps in maintaining control and clarity over one's schedule.

In essence, maximizing efficiency is about aligning daily activities with personal values, energy, and goals. By treating time like data—tracking, analyzing, and adjusting, it is possible to gain control, clarity, and calm in both professional and personal life.

9.   Screen Time and Digital Behavior

Understanding Digital Habits:

In today's technology-driven world, our daily routines are increasingly intertwined with digital devices. Every tap, swipe, and click on our screens contributes to a pattern of digital behavior that has a significant influence on our lives. Understanding these digital habits is crucial as they shape our productivity, mental well-being, and even our social interactions.

Digital habits encompass the various ways we interact with technology, from the amount of time spent on devices to the specific applications we use frequently. The data generated through these interactions is a valuable resource for analyzing our behavior. This data, often collected passively by our devices, offers insights into how we allocate our attention and time, revealing patterns that might not be immediately apparent.

One of the key aspects to consider is the amount of time spent on screens daily. Many individuals are unaware of their actual screen time, often underestimating the hours spent on phones, tablets, and computers. Most modern devices come equipped with features that track and report screen time, providing a clear picture of digital consumption.

This information can serve as a wake-up call for those who find themselves spending excessive time on non-productive activities.

Another component of digital habits is the type of applications utilized. Social media, gaming, and entertainment apps are major contributors to screen time. By analyzing app usage data, individuals can identify which applications are consuming the most time and whether this aligns with their personal and professional goals. For instance, excessive time spent on social media might indicate a need to set boundaries and prioritize more productive activities.

Digital habits are also influenced by the frequency of notifications and our response to them. Notifications are designed to capture attention, often leading to task switching and a decrease in focus. Each time a notification prompts a check-in, it disrupts concentration and can create a cycle of continuous partial attention. Being aware of this can help in managing notifications more effectively, such as by turning off non-essential alerts or setting specific times to check messages.

Task switching, or the tendency to frequently alternate between different digital tasks, is another habit that impacts productivity. Studies have shown that multitasking can reduce efficiency and increase stress, as the brain takes time to refocus after each switch. Recognizing this pattern allows individuals to design more focused work sessions, where distractions are minimized, and attention is dedicated to single tasks.

To better understand and potentially alter digital habits, individuals can engage in self-audits. This involves tracking digital usage over some time, noting not only the time spent but also the emotional and mental states experienced before and after device use. Such audits can reveal insights into which digital interactions are energizing and which are draining, enabling more informed decisions about technology use.

Primarily, understanding digital habits is about gaining awareness of how technology influences our daily lives. By analyzing the data from our digital interactions, we can make conscious choices to align our technology use with our values and goals, leading to improved productivity, better mental health, and more meaningful connections both online and offline. This proactive approach transforms digital habits from mindless routines into intentional practices that support personal and professional growth.

Balancing Online and Offline:

In the contemporary world, the line between our online and offline lives is increasingly blurred. Our daily routines often involve a seamless transition between digital interactions and physical activities, making it essential to find a balance that supports our well-being and productivity. This balance is crucial in managing the overwhelming influx of information and maintaining a healthy lifestyle.

Digital devices have become an integral part of our lives, offering conveniences that are hard to ignore. They provide instant access to information, facilitate communication, and offer entertainment. However, excessive reliance on digital tools can lead to negative consequences such as digital fatigue, decreased attention spans, and impaired social interactions. Therefore, it is essential to manage our engagement with technology consciously.

Effective management of digital and physical interactions begins with awareness. Recognizing the patterns of our digital usage can help in identifying areas that require adjustment. For instance, tracking screen time and app usage provides insights into how much time is spent on productive versus passive activities. This data can be a starting point for setting boundaries, such as limiting social media usage or designating tech-free times during the day.

Incorporating offline activities into our daily schedules is equally essential. Engaging in physical activities, face-to-face interactions, and hobbies outside the digital realm can significantly enhance our quality of life. These activities not only provide a break from screens but also encourage mindfulness and presence in the moment. For example, taking a walk, reading a book, or having a conversation without the interruption of digital notifications can be rejuvenating.

Balancing online and offline life involves setting intentional goals for digital use. This means identifying the purpose behind each digital interaction and ensuring it aligns with personal and professional objectives. By prioritizing tasks that contribute to long-term goals, individuals can optimize their digital engagement for productivity and satisfaction.

Creating a structured routine that allocates specific times for online activities can help maintain this balance. Setting times for checking emails, social media, and other digital tasks prevents them from encroaching on offline time. Additionally, utilizing digital tools like calendars and reminders can help organize tasks, ensuring that technology serves as an aid rather than a distraction.

The key to balancing online and offline life lies in the conscious use of technology. By being mindful of how, when, and why we use digital devices, we can harness their benefits without compromising our offline experiences. This balance not only enhances personal well-being but also fosters more meaningful interactions and a more fulfilling life.

Ultimately, the goal is not to eliminate technology but to integrate it thoughtfully into our lives. By doing so, we can enjoy the conveniences of the digital world while staying connected to the physical world, ensuring that neither dominates the other. This equilibrium allows us to lead a more balanced, intentional, and enriched life.

Impact on Productivity:

In the realm of modern work environments, the intersection of data analysis and productivity has emerged as a significant area of focus. The ability to leverage data to enhance productivity is not merely a theoretical exercise but a practical necessity in today's fast-paced and information-rich workplaces.

Data analysis provides a robust framework for understanding various factors that influence productivity, enabling individuals and organizations to optimize their operations and workflows.

One of the primary ways data impacts productivity is by facilitating the identification of time-wasting activities. Through the collection and analysis of time-use data, individuals can pinpoint activities that drain time without contributing to meaningful progress. This process often involves tracking daily activities, categorizing them, and analyzing patterns to identify areas for improvement. For instance, frequent social media breaks or extended meetings with little outcome can be significant productivity drains. Recognizing these patterns enables individuals to make informed decisions about eliminating or reducing such activities, thereby freeing up time for more productive tasks.

Energy management is another critical aspect of productivity that data analysis can enhance. Productivity is not solely dependent on the amount of time spent on tasks but also on the energy levels during those tasks. By analyzing data related to personal energy cycles, individuals can align their most demanding tasks with their periods of peak energy. This might mean scheduling creative or high-focus work during times when they feel most alert, such as in the morning, and reserving routine administrative tasks for when energy levels dip, typically in the afternoon.

The implementation of structured systems and routines is another way data can enhance productivity. By establishing routines, individuals reduce decision fatigue and create a predictable environment that supports sustained productivity. Systems such as time blocking, where specific periods are dedicated to particular types of work, help maintain focus and reduce the cognitive load associated with task switching. This systematic approach is bolstered by digital tools, such as calendars and task management apps, which provide the necessary structure and visibility into one's workflow.

Moreover, data empowers individuals to set boundaries based on empirical evidence. By reviewing time logs and productivity data, individuals can make informed decisions about which tasks to prioritize, delegate, or decline. This data-driven boundary setting ensures that time is allocated to activities that align with one's goals and values, fostering a more intentional and focused work environment.

Reflective practices also play a significant role in enhancing productivity through data analysis. Regular reflection on time usage, energy levels, and task outcomes allows individuals to adjust their strategies and improve continuously. Weekly reviews, for instance, provide an opportunity to assess what worked well, what caused stress, and what changes are necessary for the coming week. This iterative process of tracking, reflecting, and adjusting is akin to a feedback loop that continually fine-tunes productivity strategies.

In conclusion, the impact of data on productivity is profound and multifaceted. By harnessing data analysis, individuals and organizations can transform their approach to work, leading to increased efficiency, improved time management, and enhanced overall productivity. This not only supports personal growth but also contributes to the broader organizational objectives, making data analysis an indispensable tool in the modern productivity toolkit.

Managing Digital Consumption

In the digital age, the consumption of digital media has become an integral part of daily life, shaping how individuals interact with technology and manage their time. The ubiquitous presence of smartphones, tablets, and computers means that digital consumption is not just inevitable but also a significant component of modern living. Understanding and managing this consumption is crucial for maintaining balance and optimizing productivity.

Digital consumption encompasses the use of various media, including social media platforms, streaming services, online shopping, and news outlets. Each interaction leaves a digital footprint, a trail of data that reveals patterns in behavior, preferences, and time usage. By analyzing these patterns, individuals can gain insights into how their digital habits impact their daily lives.

One approach to managing digital consumption is through self-awareness and data tracking. Most smartphones and digital devices now provide features that track screen time, app usage, and notification frequency. These tools offer valuable data points that can be used to assess digital habits. For example, by reviewing screen time reports, individuals can identify which apps consume the most time and consider whether these align with their personal and professional goals.

The feedback loop is another important concept in managing digital consumption. This involves observing digital behavior, identifying triggers for excessive use, and setting boundaries to minimize distractions. For instance, if a person finds that they often check social media during work hours, they can set specific times for these activities or use app timers to limit usage. This method not only helps in reducing unnecessary screen time but also enhances focus and productivity.

Distinguishing between productive and passive digital use is crucial. While some digital activities, such as reading articles or using educational apps, contribute positively to learning and growth, others, like endless scrolling through social media feeds, may lead to wasted time and decreased attention spans. By categorizing digital activities into productive and passive, individuals can prioritize meaningful interactions with technology.

Emotional check-ins are also a vital part of managing digital consumption. Before, during, and after engaging with digital content, individuals should assess their emotional state. This could involve asking questions like: "Do I feel energized or drained?" "Am I more connected or isolated?" "Is this use adding value to my life?" These reflections can guide adjustments in digital habits, leading to more intentional use of technology.

Leveraging technology to achieve positive outcomes is another effective strategy. While digital devices can be a source of distraction, they can also be powerful tools for personal growth and self-improvement. Apps designed for meditation, focus, and habit-building can transform how individuals interact with their devices, turning potential distractions into opportunities for personal growth and development.

Weekly reviews of digital consumption patterns can provide ongoing insights into your usage habits. By regularly assessing digital habits, individuals can identify areas for improvement and celebrate successful adjustments. This practice encourages a proactive approach to managing digital consumption, fostering a healthier relationship with technology.

Finally, managing digital consumption is about taking control of one's digital life. It involves setting boundaries, making informed decisions, and using technology intentionally to enhance rather than hinder personal and professional goals. By treating digital consumption as a data-driven process, individuals can achieve greater balance and fulfillment in their interactions with technology.

10. Relationships and Emotional Analytics

Emotional Data Awareness:

Understanding emotions as data begins with recognizing that feelings are more than just fleeting experiences-they are significant indicators of our internal state and can be as informative as any numerical data set. Emotions provide valuable insights into our daily lives, influencing creativity, motivation, and overall well-being. By learning to recognize and track these emotional patterns, individuals can gain a deeper understanding of themselves and make informed decisions to improve their mental health and productivity.

Emotions serve as signals rather than judgments. They are not inherently good or bad, but instead, they indicate underlying conditions that require attention. For instance, anxiety may signal that something feels risky or uncertain, while excitement could indicate a new and promising opportunity. Frustration may suggest an obstacle, while joy could indicate that something aligns with personal values or desires. Analyzing these emotional signals can help identify patterns and triggers that affect mental states.

To effectively utilize emotional data, one can start by tracking one's mood. This involves noting what emotions are felt, when they occur, and the surrounding circumstances. A simple mood log can include the time, mood rating, trigger, and any additional notes. For example, feeling low in the morning might be linked to a rushed start or lack of breakfast, while feeling focused later in the day could be attributed to a quiet work environment. Over time, reviewing these logs can reveal patterns such as when mood dips typically occur, what activities boost motivation, and the emotional cost of certain habits.

Consider the case of Sudha, a graduate student who noticed her motivation waning midway through the week. By logging her emotions, she discovered that her Wednesdays were filled with back-to-back classes without breaks or rewards. By adjusting her schedule to include a short walk and lighter work on Thursdays, she achieved steadier energy and a better mood throughout the week. This example demonstrates how emotional data can be leveraged to optimize personal schedules and boost productivity.

Emotional patterns that matter include identifying triggers, such as people, environments, thoughts, or tasks, that influence one's mood. Trends, such as recurring emotional patterns over time, and thresholds, which determine the amount of input required to shift one's mood, are also essential. By understanding these patterns, individuals can achieve greater self-awareness and self-regulation.

Tracking motivation and focus is another crucial aspect of emotional data awareness. It's important to recognize

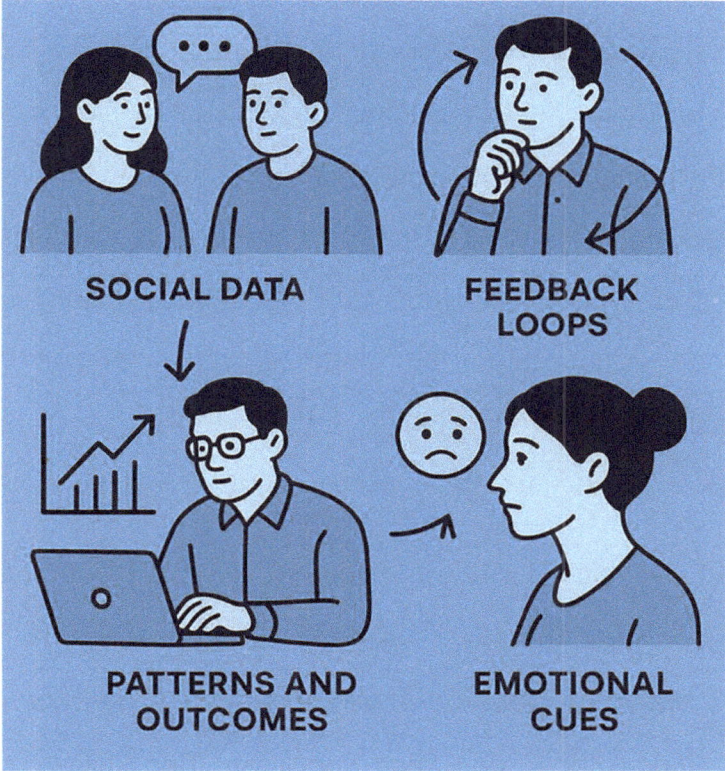

when tasks feel easy or energizing versus when they lead to procrastination. Mapping motivation against time of day or task type can reveal optimal periods for productivity and help in structuring daily activities more effectively.

In summary, emotional data awareness is about recognizing and analyzing the emotional signals that guide our lives. By paying attention to these signals, individuals can make informed decisions that enhance their emotional well-being and productivity. This approach not only aids in personal growth but also empowers individuals to lead more balanced and fulfilling lives.

Analyzing Social Interactions:

The study of social interactions presents a fascinating opportunity to explore the complexities of human communication and relationships. The essence of analyzing social interactions lies in understanding the subtle patterns and signals exchanged during conversations, which can significantly influence outcomes and relationships. By observing these interactions through a data-centric lens, we can cultivate empathy, enhance communication, and at last foster stronger connections.

Social interactions are intricate systems of data exchange and communication. Every conversation involves a multitude of data points, such as the words spoken, the tone of voice, and the accompanying body language. These elements, when combined, form a rich tapestry of information that, upon analysis, can reveal underlying patterns and dynamics within relationships. For instance, the choice of words and tone can convey confidence, uncertainty, or empathy, each eliciting different responses from the interlocutor.

To analyze social interactions effectively, one must consider the feedback loops inherent in communication. Each interaction is a cycle of action and reaction, where a statement or gesture prompts a response, which in turn influences the following action. This feedback loop can be dissected to

128

understand how specific inputs, like a supportive comment or a critical remark, affect the flow and outcome of a conversation. By adjusting these inputs, individuals can learn to steer conversations towards more positive and constructive outcomes.

Another critical aspect of analyzing social interactions is recognizing emotional data, which is often non-verbal and requires keen observation. Emotional data includes facial expressions, gestures, and postures that communicate feelings and attitudes. Understanding these cues can provide insight into the emotional state of individuals and the overall tone of the interaction. For instance, a smile or nod can indicate agreement or encouragement, while crossed arms or an averted gaze might suggest discomfort or disagreement.

Practical analysis of social interactions also involves tracking the patterns over time. Just as in any data analysis, identifying recurring themes or behaviors can help predict future interactions and outcomes. For example, if a colleague consistently responds positively to collaborative discussions but withdraws during confrontational exchanges, this pattern can inform strategies to engage them more effectively in future interactions.

Analyzing digital communication patterns has become increasingly relevant in today's technologically driven world.

Text messages, emails, and social media interactions present unique challenges and opportunities for data analysis. These digital exchanges often lack the richness of face-to-face interactions, but they still convey vital information through the timing of responses, the choice of words, and the frequency of communication. By examining these elements, one can gain insights into the digital persona and relational dynamics of individuals.

The goal of analyzing social interactions is to foster better understanding and connection. By applying data analysis techniques to human interactions, we can become more aware of our communication styles and their impact on others. This awareness enables us to adjust our approaches, leading to more meaningful and effective interactions. As we refine our ability to analyze social data, we build stronger, more empathetic relationships that are essential in both personal and professional contexts.

Building Stronger Connections:

Data analysis is not limited to numbers and spreadsheets; it extends into the realm of human interactions and relationships. Understanding how to analyze and improve our connections with others can lead to stronger, more meaningful relationships. This involves recognizing patterns, identifying areas for improvement, and making informed decisions based on the data we gather from our interactions.

One of the key elements in building stronger connections is maintaining a consistent communication frequency. Often, we may feel disconnected from someone simply because we haven't been in touch with them in a while. By tracking how frequently we engage with family, friends, or colleagues, we can identify who might require more attention. This is where relationship cadence data comes into play, revealing the frequency and timing of our interactions and helping us allocate our time more effectively.

Emotional patterns also play a significant role in strengthening connections. Reflecting on how interactions make us feel can provide valuable insights. Did the conversation boost or drain your energy? Were you actively listening, or did you dominate the conversation? These reflections serve as qualitative signals, enabling us to assess the quality of our connections and make necessary adjustments.

Conflict triggers are another area where data analysis can be beneficial. Recognizing the topics or situations that tend to cause tension allows us to adjust our approach. For instance, avoiding specific discussions during stressful times or altering our tone can help reduce relational risks. By spotting these trends, we can proactively manage conflicts and improve our interactions.

Investing in relationships is crucial for their growth and maintenance. Simple gestures, such as sending a thank-you note or checking in during difficult times, can significantly strengthen bonds. Keeping track of these actions ensures that they are done mindfully rather than mechanically, fostering intentional connection planning.

Balancing conversation dynamics is another aspect of building stronger connections. Over time, noticing who usually initiates contact, who listens more, or if there are imbalances in support can help us recalibrate our roles in relationships. Awareness of these dynamics allows for greater reciprocity and mutual support.

Digital communication patterns, such as text messages and emails, also provide valuable data. Analyzing response times, message tones, and time-of-day patterns can help us adjust our expectations and timing, leading to more effective communication.

Several tools are available to enhance relational awareness. Journaling about social interactions, setting calendar reminders to reconnect with people, or keeping notes about shared interests and milestones are small but impactful actions. These practices build relational memory and encourage deeper connections.

By approaching relationships with intention and using data to inform our interactions, we can foster stronger, more empathetic connections. While data cannot replace emotion, it can certainly enhance our awareness, responsiveness, and consideration in our relationships. Building stronger connections is about being mindful of the data our interactions generate and using it to foster more meaningful and lasting bonds.

Emotional Intelligence Development:

Several tools are available to enhance relational awareness. Journaling about social interactions, setting calendar reminders to reconnect with people, and keeping notes on shared interests and milestones are small but impactful actions. These practices help build relational memory and encourage deeper connections.

By approaching relationships with intention and using relevant information to guide our interactions, we can foster stronger and more empathetic connections. While data cannot replace emotions, it can certainly enhance our awareness, responsiveness, and consideration in our relationships. Building stronger connections involves being mindful of the information generated by our interactions and using it to create more meaningful and lasting bonds.

Emotional intelligence plays a crucial role in personal and professional development, serving as a foundational component in understanding and effectively managing emotions. Recognizing, understanding, and managing our own emotions, as well as recognizing, understanding, and influencing the feelings of others, are key skills that can be developed and refined over time. This development is not just about feeling better but about using emotional insights to make informed decisions, improve relationships, and navigate challenges more effectively.

The first step in developing emotional intelligence is self-awareness. This involves paying close attention to one's own emotions and understanding how they influence thoughts and behaviors. Keeping a mood journal can be a practical tool in this process. By recording daily emotional experiences, one can identify patterns and triggers that affect one's mood. This practice not only enhances self-awareness but also provides data that can be analyzed to identify emotional trends and their impacts on daily life.

Once self-awareness is established, the next step is self-regulation. This involves managing one's emotions, particularly in stressful situations, and maintaining control over impulsive feelings and behaviors. Techniques such as deep breathing, meditation, and mindfulness can be effective in enhancing self-regulation. These practices help in pausing before reacting, allowing for more thoughtful responses rather than impulsive reactions.

Motivation is another critical aspect of emotional intelligence. It involves utilizing emotional factors to drive self-improvement and achieve goals. Setting personal goals and tracking progress can help maintain motivation and drive. Understanding the emotions that drive personal motivation can be enlightening and lead to more effective goal setting and achievement.

Empathy, the ability to understand and share the feelings of others, is also a key component of emotional intelligence. Developing empathy involves being fully present and actively listening during interactions with others. This means paying attention not only to what is being said, but also to nonverbal cues, such as body language and tone of voice. Empathy enhances interpersonal relationships and is crucial for effective communication.

Social skills are a crucial component of emotional intelligence. This includes the ability to manage relationships and build networks. Strong social skills encompass the ability to communicate clearly, manage conflicts effectively, and collaborate effectively within a team. Practicing these skills in various social settings can enhance one's ability to interact with others in a positive and constructive manner.

Incorporating these elements into daily life requires commitment and practice. Over time, as emotional intelligence develops, individuals often find themselves better equipped to handle life's complexities. They become more adept at managing stress, communicating effectively, and building meaningful relationships. This development not only contributes to personal well-being but also enhances professional success, as emotional intelligence is increasingly recognized as a critical factor in leadership and organizational effectiveness.

By treating emotional intelligence as a form of data analysis, individuals can track their progress and identify areas for improvement. This approach encourages continuous growth and adaptation, ensuring that emotional intelligence remains a dynamic and integral part of personal and professional development.

11.   Learning from Learning

Learning isn't just about what we study; it's also about *how* we study. In this chapter, we take a closer look at the way we learn and how we can improve it by paying attention to our habits, memory, and progress. By understanding what works best for us, we can make learning more effective and easier.

We'll explore four simple ideas: how to develop better study habits, how memory aids learning, how data can inform our personal growth, and how modern educational tools can support students and teachers. When we learn from our learning, we become better at learning itself, and that helps us in school, work, and everyday life.

Study Habits Analysis

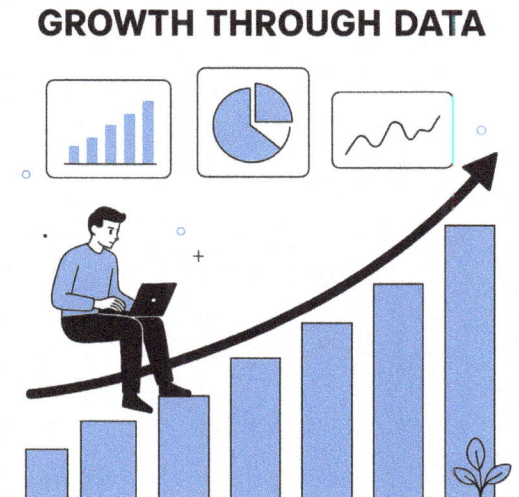

GROWTH THROUGH DATA

Understanding how we learn is a crucial aspect of improving educational outcomes. The effectiveness of study habits does not solely depend on the quantity of time spent, but rather on the quality and methods employed during study sessions. Recognizing this, it becomes essential to analyze and optimize study habits through data-driven insights.

First, we must consider the common misconception that longer study hours equate to better understanding and retention. Data suggests that the manner and timing of study sessions significantly impact learning effectiveness. For instance, studying in short, focused intervals with breaks in between, known as the Pomodoro Technique, can enhance concentration and information retention. The Pomodoro Technique involves setting a timer for a 25-minute study session, followed by a 5-minute break. After four such sessions, take a more extended break. Furthermore, the time of day can influence cognitive performance, with some individuals performing better in the morning, while others peak in the evening.

To optimize study habits, it is beneficial to track learning inputs and outcomes. A simple study log can be instrumental in this regard. By recording the date, subject, time spent, study method (such as reading, watching, or quizzing), retention the next day, and focus level, students can discern patterns and identify the most effective strategies for themselves. This personalized approach to learning allows students to adjust their study methods based on empirical evidence rather than intuition.

The concept of a learning feedback loop is pivotal. This loop involves selecting a study method, checking understanding through testing or teaching, observing what information was retained or forgotten, and adjusting the approach for future sessions. For instance, after a study session, you could test yourself on the material to see what you've retained. If you struggle with certain concepts, you can then adjust your study plan to focus more on those areas. Emphasizing retrieval practice over passive review is crucial; activities such as self-quizzing, teaching the material, or writing from memory are far more effective than merely rereading the material. This feedback loop not only enhances learning but also builds a personalized study strategy that aligns with the individual's cognitive preferences.

A practical example is Meena, a nursing student who initially struggled with low test scores despite dedicating three hours nightly to reading. By tracking her study habits, she discovered that quizzing herself led to higher retention compared to rereading. Additionally, she found that her focus was sharper in the morning, prompting her to restructure her routine to include morning study blocks and evening reviews using flashcards. This adjustment resulted in improved test scores and reduced stress, demonstrating the power of data-driven study habit modifications.

Understanding the tools and strategies that enhance retention is essential. Techniques such as spaced repetition, active recall, and teaching others are highly effective. Spaced repetition involves reviewing material at increasing intervals to reinforce memory. Active recall is the process of actively retrieving information from memory, such as through self-quizzing or other similar methods. Teaching others not only strengthens your understanding but also helps you identify areas where you have knowledge gaps. Conversely, methods like cramming and passive highlighting typically yield lower retention unless paired with active engagement strategies. By identifying and employing the most effective techniques, students can enhance their learning efficiency and effectiveness.

Ultimately, study habits are a form of personal data system. By observing time, focus, and results, students can measure and refine their learning processes to improve their outcomes. This analytical approach not only enhances academic performance but also fosters a deeper understanding of one's learning style, laying the groundwork for lifelong learning and success.

Memory Patterns

Our daily lives are filled with routines and habits that we often overlook as mere automatic responses. However, these routines are deeply rooted in memory patterns, which are essential to practical data analysis. Memory patterns refer to the way our brains store and retrieve information, allowing us to recognize and predict patterns based on past experiences. This cognitive process is akin to a natural data analysis system, where the brain collects inputs, identifies patterns, and makes informed decisions based on them.

Memory patterns are crucial for understanding how we process information in our everyday activities. For instance, when you consistently take a specific route to work, your brain remembers the landmarks, traffic patterns, and even the time it takes to reach your destination. This information is stored in your memory, allowing you to make quick decisions, such as choosing an alternative route when faced with heavy traffic. This process is not unlike how computers use algorithms to analyze data and provide solutions.

Memory patterns help us identify trends and anomalies. By recognizing repeated patterns in data, we can predict future outcomes and adjust our actions accordingly. For example, if you notice that your energy levels dip around

141

the same time every afternoon, consider changing your schedule to include a short break or a snack to boost your energy. This is a direct application of recognizing memory patterns to improve personal efficiency and productivity.

Furthermore, memory patterns are foundational in learning and adapting to new information. When learning a new skill, your brain relies on memory patterns to associate new knowledge with existing information, forming connections that enhance understanding and retention. This is why repetition and practice are vital components of skill acquisition – they reinforce memory patterns, making it easier to recall and apply what you have learned.

In professional settings, understanding memory patterns can enhance data analysis skills. Analysts use these cognitive processes to sift through vast amounts of data, identifying key patterns and trends that inform business strategies and decision-making. By harnessing the power of memory patterns, professionals can enhance their analytical capabilities, leading to more accurate predictions and practical solutions.

On a broader scale, memory patterns also play a significant role in societal and cultural contexts. Shared experiences and collective memories shape societal norms and behaviors, influencing how communities respond to events and changes.

Recognizing these patterns can provide valuable insights into cultural trends and shifts, helping to inform the development of policies and initiatives that align with community needs and values.

In conclusion, memory patterns are an integral part of our cognitive processes, influencing how we interpret and respond to the world around us. By understanding and leveraging these patterns, we can enhance our data analysis capabilities, improve personal and professional decision-making, and gain deeper insights into the dynamics of society. Embracing the role of memory patterns in our lives allows us to navigate the complexities of modern living with greater awareness and efficiency.

Growth Through Data:

Growth is not merely a byproduct but a deliberate pursuit. The ability to harness data effectively can propel individual and organizational growth by providing insights that drive informed decision-making and strategic planning. The essence of growth through data lies in transforming raw information into actionable intelligence, enabling a proactive approach to challenges and opportunities.

Data-driven growth begins with understanding the landscape of data available. This involves identifying the types of data sources relevant to your needs, such as internal metrics, customer feedback, market trends, or competitive analyses. With a clear view of these sources, one can start to gather and organize data in a way that reveals meaningful patterns and trends.

The process of data analysis involves several key steps: collecting, cleaning, and analyzing data to extract insights. Collecting data is the first step, where data is gathered from various sources. Cleaning the data is crucial, as it involves removing inaccuracies and inconsistencies to ensure the reliability of the analysis. Once cleaned, the data is analyzed using statistical methods and tools to identify patterns, correlations, and anomalies that can inform decisions.

Visualization plays a significant role in making data comprehensible. By converting complex data sets into visual formats such as graphs, charts, and dashboards, analysts can communicate insights more effectively. Visualization helps identify trends at a glance and makes data-driven arguments more persuasive.

Growth through data also involves setting measurable goals and metrics that align with strategic objectives. By defining clear key performance indicators (KPIs), organizations can track progress over time and adjust strategies accordingly. This iterative process ensures that growth is not only targeted but also sustainable.

Another critical aspect of growth through data is fostering a culture that values data-driven decision-making.

This involves training teams to interpret data insights and encouraging a mindset that prioritizes evidence over intuition. In such environments, decisions are backed by data, reducing the likelihood of errors and enhancing overall efficiency.

Data analysis enables organizations to anticipate market changes and adapt swiftly. By analyzing historical data and current trends, businesses can forecast future scenarios and prepare for potential disruptions. This foresight is invaluable in navigating competitive markets and seizing emerging opportunities.

In personal development, data can be used to track progress and optimize personal habits. Whether it's monitoring fitness levels, financial expenditures, or learning progress, personal data analysis empowers individuals to make informed decisions about their lives.

Ultimately, growth through data is about leveraging the power of information to drive continuous improvement. It involves a commitment to learning from each data point and using that knowledge to enhance decision-making processes. Whether applied to individual goals or organizational strategies, the principles of data-driven growth foster a proactive approach to achieving success. By embracing data as a fundamental tool for development, individuals and organizations can unlock new levels of potential and innovation.

Educational Data Tools:

In contemporary education, data tools have transformed the way educators, students, and administrators approach learning, making it more personalized, efficient, and impactful. These tools serve as bridges between raw data and actionable insights, allowing stakeholders to make

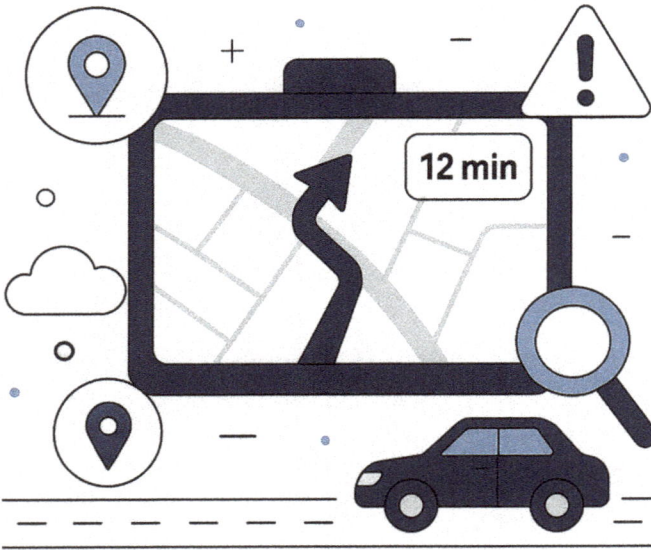

DATA-DRIVEN NAVIGATION

informed decisions that enhance educational outcomes.

One of the primary functions of educational data tools is to track and analyze student performance. These tools aggregate data from various sources such as test scores, attendance records, and behavioral reports, providing a comprehensive picture of a student's progress. By identifying trends and patterns, educators can tailor their teaching strategies to meet the diverse needs of their students. For instance, if a student consistently struggles with mathematics, data tools can highlight this trend early on, enabling teachers to intervene with targeted support or alternative teaching methods.

Also, educational data tools facilitate real-time feedback, a critical component in the learning process. Through platforms that support instant assessment, students can receive immediate feedback on their work, allowing them to understand their mistakes and correct them promptly. This instant feedback loop not only enhances learning but also keeps students engaged and motivated.

Collaboration is another area where educational data tools excel. Tools like learning management systems (LMS) and collaborative platforms allow teachers to share resources, track assignments, and communicate with students and parents seamlessly. This connectivity ensures that all parties are informed and involved in the student's learning journey, fostering a supportive educational environment.

Administrators, too, benefit significantly from educational data tools. By analyzing data on school performance, resource allocation, and staff effectiveness, administrators can make data-driven decisions that improve school operations. For example, understanding which resources are underutilized can lead to more effective budget planning and resource allocation.

Furthermore, data tools aid in predicting educational outcomes. Predictive analytics can forecast student enrollment trends, potential dropouts, and even future academic performance. These predictions help schools proactively address issues, such as increasing support for at-risk students or adjusting curriculum offerings based on projected needs.

Despite their numerous advantages, implementing educational data tools presents challenges. Data privacy and security are paramount concerns, as schools must ensure that sensitive information is protected and safeguarded. Additionally, there is a need for educators and administrators to receive adequate training on how to use these tools effectively. Without proper understanding and utilization, the potential of data tools remains untapped.

In conclusion, educational data tools are revolutionizing the education sector by making learning more personalized, efficient, and collaborative. They empower educators with insights that drive effective teaching and administrative strategies. As technology continues to evolve, the role of data tools in education is set to expand, promising even greater advancements in how we teach and learn.

12. Travel, Navigation, and Decision-Making

Planning with Precision:

Planning, in its essence, is a structured approach to achieving goals. It involves setting objectives, identifying resources, and determining the steps necessary to reach a desired outcome. In data analysis, planning is crucial because it enables the systematic collection and examination of data to inform decisions.

To begin with, effective planning requires clarity in defining the problem or question at hand. This involves understanding the scope and the specific objectives of the data analysis. Without clear goals, the process can become aimless, leading to wasted time and resources. It is essential to articulate what you are trying to achieve and why it matters. This clarity guides the subsequent stages of data collection and analysis.

Once the objectives are set, identifying the necessary data and resources becomes the next step. This involves determining what data is needed, where it can be sourced, and how it will be collected and managed. The quality and relevance of data are critical, as they directly impact the validity of the analysis. Gathering reliable data ensures that the conclusions drawn are based on accurate and comprehensive information.

Data collection methods should be selected based on the type of data required and the available resources. Whether using surveys, experiments, or secondary data sources, the technique must align with the analysis's objectives. This stage also involves considering the tools and technologies that will facilitate data collection and analysis. Selecting appropriate software or analytical tools can streamline the process and enhance accuracy.

After collecting the data, the next phase is data preparation, which includes cleaning and organizing the data for analysis. This step involves checking for errors, inconsistencies, and missing values, which can skew results if not addressed. Data preparation ensures that the dataset is ready for analysis and that the results will be reliable.

With a clean dataset, analysts can proceed to data analysis, applying statistical techniques and models to uncover patterns, trends, and relationships. This stage requires a keen understanding of the data and the questions being addressed. Analysts must choose the proper analytical methods to extract meaningful insights from the data.

Throughout the planning process, it is vital to maintain flexibility. As analysis progresses, new insights or challenges may arise, necessitating adjustments to the original plan. This adaptability ensures that the analysis remains relevant and aligned with the objectives.

Communicating the results of the analysis is a crucial part of the planning process. This involves presenting the findings in a clear and accessible manner, tailored to the audience's needs. Effective communication ensures that the insights gained from the analysis are understood and can inform decision-making.

In conclusion, planning with precision in data analysis involves setting clear objectives, collecting and preparing data meticulously, employing appropriate analytical methods, and communicating results effectively. By following a structured planning process, analysts can ensure their work is focused, efficient, and impactful, primarily leading to better decision-making and outcomes.

Data-Driven Navigation:

Navigation is a prime example of how data-driven decision-making can simplify and enhance everyday experiences. Whether commuting to work, embarking on a road trip, or simply finding the quickest way to a new restaurant, navigation relies heavily on data collection, pattern recognition, and predictive analytics.

At the heart of this process is the collection of real-time data. Modern navigation systems, such as GPS and apps like Google Maps and Waze, continuously gather data from numerous sources, including satellite signals, user inputs, and traffic updates. This wealth of information allows these systems to assess current road conditions, predict traffic patterns, and suggest the most efficient routes. This dynamic use of data exemplifies how real-time analytics can lead to optimal decision-making in navigating complex environments.

Pattern recognition plays a critical role in navigation. By analyzing historical data, navigation systems can identify traffic trends, such as peak congestion times and common bottlenecks. This historical insight enables them to forecast future conditions and adjust route suggestions accordingly. For example, if a particular highway is known for heavy traffic during rush hour, the system can preemptively suggest alternate routes to avoid delays.

Predictive analytics further enhances navigation by anticipating potential disruptions and adjusting recommendations in real time. By integrating weather forecasts, construction schedules, and accident reports, navigation systems can alert users to potential hazards and provide alternative routes. This proactive approach not only saves time but also enhances safety by minimizing exposure to adverse conditions.

The integration of user-generated data is another vital

aspect of data-driven navigation. Many navigation apps enable users to report real-time incidents, including accidents, road closures, and speed traps. This crowd-sourced data enriches the system's database, providing a more comprehensive view of current conditions. As more users contribute information, the accuracy and reliability of the navigation system improve, creating a collaborative ecosystem of data sharing.

Personalized navigation experiences are becoming increasingly common. By analyzing individual user data, such as past travel history, preferences for specific types of roads, or avoidance of tolls, navigation systems can tailor their recommendations to meet individual needs. This personalization not only enhances user satisfaction but also demonstrates the power of data in creating customized solutions.

Data-driven navigation also extends beyond personal vehicles. Public transportation systems utilize data analytics to optimize schedules, manage fleet operations, and improve passenger experiences. By analyzing ridership patterns, transit authorities can adjust service frequencies, deploy resources more efficiently, and enhance route planning. Similarly, logistics companies utilize data to streamline their delivery operations, thereby reducing costs and improving service reliability.

Data-driven navigation exemplifies the transformative power of data in everyday life. By leveraging real-time data, pattern recognition, predictive analytics, and user-generated insights, navigation systems provide efficient, safe, and personalized travel experiences. As technology continues to advance, the integration of data into navigation will only deepen, offering even greater benefits and innovations in how we move through the world.

Evaluating Travel Options:

When selecting the best travel options, several factors come into play, each requiring careful consideration to ensure a balanced decision. Firstly, cost analysis is a primary concern for most travelers. This includes not just the price of tickets or fuel, but also the hidden costs associated with each mode of transport. For instance, opting for a flight might seem costly at first glance, but when factoring in the time saved and potential conveniences, such as included meals or amenities, the value proposition might shift in its favor.

Another critical aspect is time efficiency. Evaluating the duration of each travel option, including potential delays and layovers, is crucial. For instance, a direct flight might be preferable over a cheaper option with multiple stops, especially when time is a constraint. Similarly, train travel can offer a unique balance of speed and scenery, often allowing travelers to work or relax more comfortably than in a car or airplane.

The environmental impact of travel choices also plays a significant role, particularly for those who are conscious of their carbon footprint. Public transportation, such as trains or buses, generally has a lower environmental impact compared to flying or driving alone. In this context, the decision-making process might involve weighing personal convenience against ecological responsibility.

Comfort and convenience are equally important. The experience of travel itself can be a deciding factor. Factors such as legroom, the ability to move around, access to food and restrooms, and even the view can influence the choice. For long journeys, the ability to sleep comfortably or have entertainment options can significantly affect the travel experience.

Safety and reliability are other crucial variables. Travelers often consider the safety record of airlines, the reliability of train schedules, or the likelihood of road congestion when planning their trips. Ensuring that the chosen mode of transportation is not only safe but also dependable can prevent unnecessary stress and uncertainty.

Furthermore, the flexibility offered by each travel option can influence decisions. Flexible tickets or the ability to change plans without significant penalties can be invaluable, especially in times of uncertainty or when traveling with family. This flexibility can sometimes justify a higher upfront cost if it means reduced stress or the ability to easily adapt plans.

Lastly, personal preferences and past experiences shape travel decisions. Someone who enjoys the scenic route might prefer driving despite the longer travel time, while others might prioritize speed and efficiency. Personal experiences, such as a previous enjoyable train journey or a stressful flight, can heavily influence future choices.

In summary, evaluating travel options is a multifaceted process that involves balancing cost, time, environmental impact, comfort, safety, flexibility, and personal preferences. By systematically analyzing these factors, travelers can make informed decisions that align with their priorities and enhance their overall travel experience.

Decision-Making on the Move:

In our daily lives, decision-making often extends beyond stationary environments. Whether it's navigating through bustling city streets or planning a cross-country trip, data plays a pivotal role in guiding our choices.

The modern traveler, equipped with an array of digital tools, continuously interacts with data to make informed decisions on the go.

Consider the daily commute, a routine that millions of people worldwide engage in. Applications like Google Maps and Waze have revolutionized the way we approach travel by providing real-time data on traffic conditions, estimated travel times, and alternative routes. These applications utilize crowd-sourced information to provide users with the most efficient paths, helping them avoid congestion and save time. Even without such tools, individuals subconsciously apply data-driven decision-making by leaving earlier due to weather forecasts, choosing back roads during peak hours, or monitoring fuel efficiency over time.

The process of selecting a mode of transportation is another area where data significantly influences decisions. Travelers must balance multiple variables, such as cost, duration, convenience, and environmental impact. For example, opting between driving, taking a bus, or cycling involves considering not only the immediate financial cost but also the time investment and ecological footprint. This multi-variable decision-making process highlights how travelers prioritize different factors based on available data.

Budgeting for travel involves expenses, a critical aspect of effective travel planning. Whether planning a weekend getaway or an extended vacation, travelers estimate costs for lodging, transportation, and food. By analyzing historical spending patterns from previous trips, individuals can make informed decisions to optimize their travel budgets and avoid unexpected expenses.

Packing for a trip is another area where data analysis plays a role. Over time, travelers learn what items are essential based on past experiences and the expected weather conditions of their destinations. This pattern analysis reduces stress and increases efficiency by ensuring that travelers are well-prepared without overpacking.

Tracking itineraries has become more streamlined with digital tools like TripIt and Google Calendar, which help travelers organize and visualize their plans. These tools enable users to consolidate all travel-related information in one place, thereby reducing cognitive load and facilitating more effective decision-making.

Ultimately, decision-making on the move is about applying data analysis to optimize travel experiences. By leveraging available data, including real-time traffic updates and historical spending patterns, travelers can make more informed and more intelligent choices. This approach not only enhances the travel experience but also empowers individuals to navigate their journeys with confidence and clarity. As data becomes increasingly integrated into our daily routines, the ability to analyze and apply this information is essential for making informed decisions on the move.

13. Community and Civic Data

We often think of data as something used by businesses or scientists, but it also plays a powerful role in shaping our communities. From public schools and hospitals to traffic systems and city planning, data is quietly working behind the scenes to improve the services we rely on every day. This type of data, often referred to as civic data, is the information collected and used by government entities, nonprofit organizations, or other public institutions to make decisions that affect our daily lives.

In this chapter, we'll explore how public data is utilized to enhance education, health, safety, and urban life. We'll also explore how everyday people like you and me can use data to get involved, raise their voices, and ensure their communities move in the right direction.

Understanding civic data is not just about knowledge; it's about empowerment. It's about being informed, engaged, and part of positive change. In this chapter, we'll explore how data enables government and civic organizations to make informed decisions, promotes transparency and accountability, and empowers us to shape the world around us actively. Most importantly, we'll see how understanding civic data allows individuals to be informed, engaged, and part of positive change.

Public Data Utilization:

Data plays a crucial role in shaping modern society, and its utilization in public domains has become a cornerstone for enhancing community welfare. Public data, which refers to information collected by government entities, nonprofit organizations, or other public institutions that is made accessible to the general public, is the backbone of our societal structure. This data encompasses a diverse range of categories, including demographic statistics, economic indicators, health records, and environmental data. By leveraging this wealth of information, various sectors can significantly improve their decision-making processes, resource allocation, and service delivery, drive to more informed and efficient public administration.

One of the primary applications of public data is in the field of education. Educational institutions utilize data to track student performance, attendance, and graduation rates. This information enables educators to identify areas where students may require additional support, allowing schools to allocate resources more effectively. For instance, analyzing test scores and attendance records can reveal trends that necessitate changes in teaching strategies or highlight the need for targeted interventions for students who are struggling. Furthermore, parents and guardians can use this data to engage more deeply with their children's education, making informed decisions about their learning and development.

In healthcare, public data is not just a tool; it's a lifeline. It plays a pivotal role in monitoring and improving public health outcomes. Hospitals and clinics collect and analyze data on vaccination rates, disease outbreaks, and patient outcomes to enhance their services. During health crises, such as the COVID-19 pandemic, real-time data on infection rates, hospital capacities, and vaccination distributions have been vital in managing the crisis effectively. This data-driven approach enables healthcare providers to respond swiftly to emerging health threats, optimize resource allocation, and improve preventive care measures.

Urban planning and development also benefit significantly from the utilization of public data. Cities use data to manage transportation systems, energy consumption, and public safety. Traffic data helps optimize traffic light timings and reduce congestion, while energy usage patterns can inform sustainable development initiatives. Additionally, public feedback collected through various channels can inform city planners' decisions about infrastructure projects and community services. By analyzing this data, cities can create safer, more efficient environments that cater to the needs of their residents.

Public data also fosters civic engagement and transparency. Citizens can access government data to understand how public funds are being used, participate in decision-making processes, and hold public officials accountable. For instance, understanding budget allocations can help citizens advocate for improved public services in their communities. This transparency strengthens democratic processes by empowering citizens with the information needed to make informed decisions and advocate for their communities.

However, the effective utilization of public data requires careful consideration of privacy and ethical concerns. Ensuring that data is anonymized and securely stored is crucial to protect individual privacy while maintaining public trust. Moreover, efforts must be made to ensure that data is accessible and understandable to all segments of the population, preventing disparities in who can benefit from data-driven insights.

In summary, the effective utilization of public data is a powerful tool for enhancing public services and promoting community well-being. By harnessing the potential of data, governments and organizations can enhance transparency, accountability, and efficiency, ultimately steering to more informed decision-making and better outcomes for society as a whole.

For instance, data can help identify areas for improvement in public services, direct to more effective resource allocation, and better service delivery.

Civic Engagement through Data:

Data is not only a tool for personal decision-making but also a powerful resource for civic engagement. It plays a crucial role in how communities function and thrive, influencing everything from education and healthcare to urban planning and governance. By understanding and leveraging data, individuals can actively participate in shaping their communities, ensuring that public services are responsive and effective.

In the realm of education, data serves as a vital instrument in tracking and enhancing student success. Schools utilize data to monitor student attendance, assess test scores, and analyze graduation rates, enabling educators to adjust teaching strategies and allocate resources effectively. This data-driven approach allows for targeted interventions, ensuring that students who are struggling receive the necessary support. Parents, too, can engage with educational data through report cards and progress reports, using this information to support their children's learning at home.

Healthcare is another sector where data proves invaluable. Health systems and public health departments gather data on vaccination rates, disease outbreaks, and patient outcomes to optimize care delivery and respond to health crises. During the COVID-19 pandemic, real-time data tracking of case numbers and vaccine distribution became a lifeline. Individuals contribute to this data ecosystem by logging health information and participating in health surveys, thus playing a part in the community's overall health management.

Urban environments rely heavily on data for planning and improvement. Traffic sensors, energy usage data, and citizen feedback apps provide cities with the information needed to manage transportation, utilities, and safety. This data helps urban planners to design better public transit systems, ensure safer neighborhoods, and direct resources where they are most needed. Public participation in data collection, such as reporting maintenance issues or filling out surveys, enhances the ability of city officials to make informed decisions.

Civic participation extends beyond just data collection; it is about making your voice heard. Whether through participating in school board meetings, filling out the census, or voting in local elections, these actions generate data that informs decisions about funding, programs, and leadership. This is democracy in action, showcasing how individual input can shape community directions and priorities.

Transparency and trust are built when data is openly shared with the public. Schools release report cards, health departments publish dashboards, and city budgets are made available for review. When citizens have access to this data, they can hold institutions accountable and advocate more effectively for their needs. Understanding these data systems enables individuals to navigate public services more effectively, thereby contributing to the overall well-being of the community.

Ultimately, data is a bridge between individual actions and collective well-being. By engaging with data, citizens can influence the decisions that impact their communities, ensuring that public services reflect the needs and desires of the people they serve. This engagement fosters a more connected, informed, and empowered society.

Improving Community Services:

Public services have increasingly relied on data to enhance community life. By systematically gathering and analyzing information, these services can allocate resources more effectively, identify needs more accurately, and deliver better outcomes for citizens. This approach is not only beneficial for institutions but also highlights the crucial role of individuals in engaging with and contributing to data-driven improvements in their communities.

In educational settings, data plays a crucial role in monitoring student progress and success. Schools utilize attendance records, test scores, and graduation rates to track and enhance student engagement and performance. Teachers and administrators analyze this data to adjust teaching

WORLD POPULATION ON SOCIAL MEDIA

64%
of the world's population uses social media

DATA THAT IS UNSTRUCTURED

80%
of all data is unstructured

DATA ACTUALLY ANALYZED (USEFUL)

1%
of data is analyzed

strategies, allocate resources efficiently, and design targeted interventions for students who require additional support. Parents also play a significant role, benefiting from accessing data such as report cards and progress reports, which enable them to actively support their children's learning at home.

Healthcare systems are another area where data plays a pivotal role. Public health institutions track vaccination rates, disease outbreaks, and patient wait times to respond swiftly to health crises, optimize staffing, and improve preventive care. The COVID-19 pandemic underscored the crucial role of data in tracking case numbers and efficiently managing healthcare resources. At the individual level, people play a vital role in contributing to health data by sharing symptoms, logging health information into apps, and participating in surveys, thereby enhancing the collective understanding of public health needs.

The collection and analysis of data heavily influence urban planning in cities. Traffic sensors, for example, adjust light signals to reduce congestion, while energy usage patterns inform sustainable initiatives. Through citizen feedback apps, residents can report maintenance issues, enabling cities to plan better public transit systems, enhance neighborhood safety, and target improvements where they are most needed. This participatory approach ensures that urban development is more aligned with the actual needs and experiences of residents.

Civic participation is enhanced through data-driven decision-making. Actions such as attending school board meetings, completing the census, and voting in local elections contribute valuable data that informs decisions about funding, programs, and leadership. Transparency in sharing data, including school report cards and city budgets, builds trust and accountability, allowing citizens to engage with and influence the institutions that serve them.

Overall, data analysis is a powerful tool for improving community services. It transforms personal concerns into compelling evidence that can drive change and shape more responsive institutions. By understanding and participating in these data-driven processes, individuals can navigate public services more effectively, advocate for their needs, and contribute to the overall well-being of their communities. Through education, healthcare, urban planning, and civic engagement, data analysis empowers communities to make informed decisions that enhance the quality of life for all citizens.

Future of Civic Data:

Cities have always been epicenters of human activity, and as they continue to grow, the role of data in managing urban life becomes increasingly significant. Civic data refers to the information collected by government entities and public organizations to enhance various aspects of city life, including transportation, public safety, utilities, and citizen engagement. The future of civic data is poised to transform urban living through more innovative, more efficient, and more inclusive systems. This could mean more intelligent traffic management, more effective crime prevention, and better resource allocation, among other benefits.

One of the primary applications of civic data is in transportation management. Modern cities utilize traffic sensors and GPS data to monitor and manage the flow of vehicles. Real-time data enables dynamic adjustments to traffic signals, thereby reducing congestion and improving commute times. Furthermore, public transit systems are increasingly data-driven, using passenger data to optimize routes and schedules, thereby enhancing efficiency and rider satisfaction.

In the realm of public safety, data analytics is being leveraged to predict and prevent crime. Law enforcement agencies utilize historical crime data to identify patterns and deploy resources more effectively. Predictive policing, although controversial, aims to allocate patrols to areas where crime is statistically more likely to occur, thereby potentially reducing incidents and improving community safety. This data-driven approach ensures that public safety remains a top priority, making citizens feel more secure and protected.

Utilities management is another area where civic data is making a significant impact. Smart meters and sensors provide real-time data on energy and water usage, enabling cities to manage resources more sustainably. This data not only helps reduce waste and lower costs but also supports the development of green initiatives by identifying areas where efficiency can be improved.

Civic data is also transforming citizen engagement and participation. Tools such as feedback apps and online platforms enable residents to report issues, participate in local governance, and provide input on community projects. This data is invaluable for city planners and officials as it reflects the community's needs and priorities, allowing for more responsive and democratic governance.

Transparency and accountability in government operations are enhanced through the open sharing of civic data. Public access to data on school performance, health department reports, and city budgets allows citizens to stay informed and hold their governments accountable. This transparency builds trust and encourages more active participation in civic life.

However, the future of civic data is not without challenges. Privacy concerns are paramount as cities collect vast amounts of data from their residents. Ensuring data security and protecting individuals' privacy rights are crucial issues that must be addressed. Additionally, there is a need for policies that ensure equitable access to data and technology, preventing the digital divide from widening further. These challenges can be addressed through robust data protection laws, ethical data use guidelines, and initiatives aimed at bridging the digital divide.

The potential for civic data to improve urban living is

immense, but it requires a careful balance between innovation and regulation. By fostering a culture of data literacy and ethical use, cities can harness the power of data to create more livable, equitable, and sustainable environments. As technology continues to advance, the integration of civic data into the fabric of city life will likely become even more seamless, driving the next wave of urban transformation forward. Individuals must adapt to this changing landscape and understand the importance of data literacy and ethical use.

14. Conclusion

The Everyday Analyst Mindset:

Throughout this book, we've explored how data isn't just something hidden in spreadsheets or complex systems; it's woven into the rhythms of our daily lives. From morning routines to travel plans, from emotional reflections to screen time habits, data provides valuable insights into what's working, what's not, and where we can make improvements. Thinking like a data analyst doesn't mean becoming technical; it means becoming more observant, curious, and intentional. By tracking, reflecting, and adjusting based on what we notice, we take small but meaningful steps toward better choices. Whether you're reviewing your week, planning your meals, or deciding how to spend your time, your life already gives you feedback. As you begin to listen, patterns emerge, clarity improves, and confidence grows.

That's the everyday analyst mindset, not perfection, but awareness. Not control, but conscious choice. As you move forward, let this mindset guide you gently: observe often, reflect honestly, and live with greater understanding. Your data isn't just numbers. It's your story, and you are its most crucial reader. See below for the job opportunity available for Data Analysts.

This book is the first in a series. It serves as a warm-up, preliminary, and introductory book. I am writing more books on Data Analysis and continue to look for their releases.

Data Analysts Wanted...:

As we approach the final stage of this journey into data analysis, it's a good time to pause and reflect on just how powerful data can be, not just in business or technology, but in the everyday moments of life.

Throughout this book, we've seen that data isn't just rows in a spreadsheet or complex algorithms on a server. It's the small decisions we make every day: choosing what to wear based on the forecast, planning dinner based on what's left in the pantry, or managing a family budget with thoughtful spending. These are data decisions, and you've been analyzing them all along, whether you realized it or not.

This infographic reveals that while 64% of the world's population uses social media and 80% of global data is unstructured, only 1% of it is analyzed. This reveals a vast amount of untapped information and underscores the rising demand for skilled data analysts. As data continues to grow, so does the need for more effective tools and expertise to transform raw information into meaningful and actionable insights.

What makes data analysis truly valuable is the mindset it builds. It teaches us to ask, "What can I learn from this?" and "How can I make this better?" That mindset turns routine experiences into opportunities for growth and learning. With every observation, every pattern, every question, we move closer to clarity and better outcomes.

And here's the exciting part: the world needs people who think this way.

As data becomes more central to decision-making in every field, from healthcare to education, finance to urban planning, the demand for skilled data analysts is growing faster than ever. Whether you're just starting or looking to pivot into a new career, there's a place for you in this expanding field. Your skills are not only valuable but also in high demand, making you a sought-after professional in today's data-driven world.

Below is a breakdown of the many types of Data Analyst roles you'll find across industries. Now that you've built a foundation, you can begin exploring which domain speaks to you and what specific skills you'll want to develop next.

General & Business-Oriented Roles:

- Business Data Analyst

- Operations Data Analyst

- Marketing Data Analyst

- Sales Data Analyst

- Financial Data Analyst

- Product Data Analyst

- Customer Insights Analyst

- Supply Chain Data Analyst

- Retail Data Analyst

- Revenue Analyst

Technology & Advanced Analytics Roles

- Data Science Analyst

- Machine Learning Data Analyst

- Data Visualization Analyst

- Web Analytics/Data Analyst

- SQL Data Analyst

- ETL/Data Integration Analyst

- Big Data Analyst

Industry-Specific Roles

- Healthcare Data Analyst
- Clinical Data Analyst
- Pharmaceutical Data Analyst
- Insurance Data Analyst
- Education Data Analyst
- Telecom Data Analyst
- Banking/Investment Data Analyst
- Energy/Utilities Data Analyst

Governance, Risk, and Compliance

- Risk Data Analyst
- Compliance Data Analyst
- Fraud/Data Security Analyst
- Data Quality Analyst
- Data Governance Analyst

Emerging or Specialized Roles

- HR/People Analytics Analyst
- Geospatial/GIS Data Analyst
- Sports Performance Data Analyst
- Environmental Data Analyst
- Media/Streaming Data Analyst
- AI Operations Data Analyst
- Behavioral Data Analyst

By now, you've learned to see the world through a more analytical lens to observe more closely, to ask better questions, and to trust the stories hidden in data. Whether you're managing your health, planning your finances, exploring your community, or considering your career path, data analysis gives you a way to approach life with intention and confidence.

As you turn the final page, remember: every moment is a data point. Every decision is a chance to learn. Keep that mindset with you, and you won't just be analyzing data; you'll be designing a life that reflects your values, goals, and growth.

Stay curious. Keep asking questions. Let the data of your life lead you forward, look for further data analysis books, keep learning, and take up one of the data analyst roles listed above.

Bibliography

1. Clear, J. (2018). *Atomic Habits: An easy & proven way to build good habits & break bad ones.* Avery.

2. Gilbert, E. (2015). Big Magic: Creative living beyond fear. Riverhead Books.

3. Harford, T. (2021). The Data Detective: Ten easy rules to make sense of statistics. Riverhead Books.

4. Knaflic, C. N. (2015). Storytelling with Data: A data visualization guide for business professionals. Wiley.

5. Newport, C. (2016). Deep Work: Rules for focused success in a distracted world. Grand Central Publishing.

6. Sinek, S. (2009). Start with Why: How great leaders inspire everyone to take action. Portfolio.

7. Wheelan, C. (2013). Naked Statistics: Stripping the dread from the data. W. W. Norton & Company.

About the Author

Mourougavelou Vaithianathan, also known as "Puduvai Murugu," is a seasoned data architect, engineer, and technologist with over two decades of global experience across India, Canada, and the United States. Born in Puducherry, India, Murugu earned his Bachelor's degree in Mechanical Engineering from Pondicherry Engineering College and later completed his Master's in Engineering from Ryerson University in Toronto, Canada.

Murugu has held diverse roles including Mechanical Engineer, Data Analyst, Data Scientist, IT Specialist, and Business Consultant. He is widely recognized for his ability to modernize legacy systems into scalable, secure, and cost-effective cloud-native data platforms.

With deep expertise in Lakehouse architecture design, Murugu leverages platforms such as Databricks, Apache Spark, and Snowflake to build advanced analytics and AI/ML solutions. He is proficient in working across multi-cloud environments (AWS, Azure, GCP) and specializes in designing decoupled storage-compute models, cloud-native ETL frameworks, and metadata-driven data ingestion pipelines. He leads cloud BI modernization, API-driven integration, and DevOps-enabled delivery models, ensuring secure, agile, and compliant enterprise data platforms.

Murugu's passion for data and continuous learning is reflected in his extensive certification portfolio, which includes:

- Snowflake Certified SnowPro Advanced Architect
- Snowflake Certified SnowPro Core Professional
- Databricks Certified Data Engineer Professional
- Databricks Certified Data Engineer
- Databricks Certified Data Analyst Associate
- Certified SAFe 5 Agilist
- Certified SAFe 5 Practitioner
- Management Excellence at Microsoft: Model, Coach, Care
- Certified Splunk Core Power User
- TOGAF 9 Training (Levels 1 and 2) – Enterprise Architecture
- SAS Certified Advanced Programmer for SAS 9
- SAS Certified Base Programmer for SAS 9

Beyond his technical accomplishments, Murugu is a dedicated linguistics enthusiast and a passionate advocate for Bilingual writing in the Thamizh language, literature, and culture. He also actively volunteers with the Fun Cycle Riders Organization, participating in group riding activities on bicycles.